食品生物工艺专业改革创新教材系列

审定委员会

主　任　周发茂

副主任　余世明

成　员（以姓氏笔画为序）

　　　　　王　刚　刘海丹　刘伟玲　许映花　许耀荣

　　　　　余世明　陈明瞭　罗克宁　周发茂　胡宏佳

　　　　　黄清文　潘　婷　戴杰卿

食品生物工艺专业改革创新教材系列　　总主编　周发茂

烘焙专业英语
English for Baking

主编 ◎ 陈明瞭

暨南大学出版社

中国·广州

食品生物工艺专业改革创新教材系列

编写委员会

总主编 周发茂

委　员（以姓氏笔画为序）

王　刚　　王建金　　区敏红　　邓宇兵　　龙伟彦
龙小清　　冯钊麟　　刘海丹　　刘　洋　　江永丰
许映花　　麦明隆　　杨月通　　利志刚　　何广洪
何婉宜　　何玉珍　　何志伟　　余世明　　陈明瞭
陈柔豪　　欧玉蓉　　周发茂　　周璐艳　　郑慧敏
胡源媛　　胡兆波　　钟细娥　　凌红妹　　黄永达
章佳妮　　曾丽芬　　蔡　阳

编写说明

本书系食品生物工艺专业（烘焙食品方向）的烘焙专业英语学生用书，是职业教育改革创新教材系列中的一本。

全书以任务为载体、以情境创设为氛围进行课程设计，重新构建课程结构，以适应中西点专业学生的英语程度。同时紧密结合烘焙食品的专业课程和操作产品，融教与学于一体，体现实践性教学的理念。

通过学习本课程，学生能够掌握一定数量的中西点制作专业英语词汇；基本掌握常用的一些短语；能说、会听常用的产品、原料、工艺过程专业术语，了解其基本应用；借助工具书能基本读懂配方，同时为今后学生报考高级工证做准备。

本书也适合烘焙食品从业人员作烘焙英语培训教材。

本书由广东省贸易职业技术学校高级讲师陈明瞭主编，参与编写人员有：龙小清（副主编、广东省贸易职业技术学校讲师）、章佳妮（广东省贸易职业技术学校高级讲师）、龙伟彦（广州白云国际会议中心副总经理、高级技师）、冯钊麟（广州市花园酒店西饼房厨师长、高级技师）。具体分工为：章佳妮负责模块一，龙小清负责模块二，陈明瞭负责模块三（其中部分由龙小清编写），龙伟彦、冯钊麟提供素材及编写部分词汇，全书由陈明瞭负责总纂。

全书的卡通形象由广东省贸易职业技术学校动漫教研组吕建雄老师、吴颖敏老师绘制，在此一并致谢！

编写说明 ·· 1

Module 1: In a Bakery and Bread Factory

Project 1 In a Bakery ·· 3
Project 2 Ordering a Birthday Cake ······················· 10
Project 3 Visiting a Bread Factory ························ 17
Project 4 Visiting a Cake Shop ······························· 24

Module 2: Formulas and Making

Project 5 Bread's Formula ······································· 35
Project 6 Cake's Formula ··· 45
Project 7 Butter Cookies ·· 52
Project 8 Some Bread Ingredients ·························· 58

Module 3: The Process of Bread Production and Dough Processing Methods

Project 9 Dough Development During Mixing ········ 67
Project 10 Chiffon Cakes ·· 73
Project 11 Dough Processing Methods ···················· 78

Project 12 Baking Reaction ………………………………………………… 85

附录一：生词和词组汇总 ………………………………………………… 91
附录二：西式面点师职业资格证书理论考试专业英文词汇 …………… 104
附录三：拓展阅读参考译文 ……………………………………………… 106

参考资料 …………………………………………………………………… 114

Module 1: In a Bakery and Bread Factory

模块一

在西饼店和面包厂

1. IN A BAKERY

4. VISITING A CAKE SHOP

YOU WILL LEARN 4 DIALOGUES.

2. ORDERING A BIRTHDAY CAKE

3. VISITING A BREAD FACTORY

Project 1
In a Bakery

Task 1 任务一：
在面包店里，店员应该如何招呼进店的顾客？用英语怎样说？（采用小组讨论形式，最后用英语在讲台上表达）

Task 2 任务二：
店员如何向顾客介绍店里的面包产品？假如用英语你怎样说？（采用小组讨论形式，最后用英语在讲台上表达）

Task 3 任务三：
用英语写出你知道的面包产品名称。（采用小组讨论形式，最后在黑板上写出，写得最多的为本节课的优胜组）

Task 4 任务四：
顾客选好面包产品后，店员如何为顾客计算金额并收款？假如用英语你怎样说？（采用小组讨论形式，最后用英语在讲台上表达）

Which do you like most?

whole wheat bread   croissant   rye bread

doughnut   sandwich   farmer bread

French bread   Danish pastry   puff pastry

puff   toast bread   hot dog

Dialogue

Mary goes to a bakery to buy bread.

M: Mary   **S**: Shop Assistant

S: Hello. What can I do for you?

M: I'd like to buy some bread. Can you recommend some for me?

S: Yes, of course. We have French bread, farmer bread, whole wheat bread, rye bread, and so on. Which do you like?

M: Could you please tell me the differences between whole wheat bread and rye bread?

Let's listen to the dialogue.

S: Yes, whole wheat bread provides many health benefits, such as protection from heart disease, stroke, and some cancers. Rye bread is particularly high in fiber and low in fat.

M: Thank you for your recommendation. Then, I'd like to choose some whole wheat bread.

(Several minutes later.)

M: How much should I pay for them?

S: Altogether 13.80 dollars, please.

M: Here is a 20 dollars note.

S: Here is your change.

M: Thank you.

Notes

我们来看看下面的注释。

1. What can I do for you?
   我可以帮助您吗？/我能为您做什么？
   一般为售货员见到顾客后说的话。
   类似的表达有：Can I help you? 和 May I help you?

2. I'd like to buy some bread.
   我想买一些面包。
   这个句型用来表达自己想要做什么。

3. Which do you like?
   您喜欢哪一种？

4. Thank you for your recommendation.
   谢谢你的介绍。

5. Here is your change.
   这是找给您的零钱。

快来学单词啰！

New Words and Expressions

1. bakery ['beikəri] n. 面包房，面包店
2. wheat [wiːt] n. 小麦
3. croissant [krwɑ'sɑː] n. 牛角酥，牛角包
4. doughnut ['dəʊnʌt] n. 油炸面包圈
5. puff [pʌf] n. 泡芙（奶油空心饼）
6. bread [bred] n. 面包
7. pastry ['peistri] n. 糕点，油酥糕点

8. hot dog ［hɔt dɔg］热狗

9. toast bread ［təust bred］吐司面包

10. French bread ［frentʃ bred］法式面包

11. Danish pastry ［ˈdeiniʃ ˈpeistri］丹麦包，丹麦酥

12. puff pastry 松饼，层酥点心，清酥

13. farmer bread ［ˈfɑːmə bred］农夫包

14. rye bread 裸麦面包

　　rye ［rai］n. 黑麦，黑麦粒

15. whole wheat bread 全麦面包

16. recommend ［ˌrekəˈmend］vt. 推荐，介绍

17. difference ［ˈdifərəns］n. 差别，差异

18. sandwich ［ˈsænwidʒ］n. 三明治，夹心面包

19. change ［tʃeindʒ］n. 变化；找回的零钱（用为不可数名词，不能加 s）

20. and so on 等等；诸如此类的；依此类推（用在诸多列举项目之后，相当于 etc.）

21. How much… ……多少钱？

22. I'd like to… 我想……

23. Thank you for… 谢谢你的……

Grammar

1. I'd like to 的相关用法

   I'd like to… 意思是"我想要……"，一般用在表示"自己想要什么、想做什么"的客气话里。例如：

   I'd like to eat some apples.
   我想吃些苹果。

   I'd like to play football.
   我想去踢足球。

   I'd like to go to Beijing.
   我想去北京。

   记住：I'd like to… 后面要用动词原形。类似的表达有：I want to…

2. "最喜欢"的表达

   What bread do you like most? Why?
   你最喜欢什么面包？为什么？

   类似的表达有：

   What's your favorite bread?

   Which bread do you like best?

   Which… do you like best ∕ most?

   What's your favorite…

In a Bakery Project ❶ 项目一

Exercises

❶ "想一想"
我都做过哪些面包？能用英语说出来吗？

❷ "试一试"
请找出与中文对应的实物，将序号写在相应的中文后面。

A. 油炸面包圈　　B. 三明治　　C. 吐司面包　　D. 农夫包　　E. 全麦面包
F. 丹麦包　　　　G. 牛角包　　H. 裸麦面包　　I. 法式面包　J. 松饼
K. 泡芙　　　　　L. 热狗

①whole wheat bread

②croissant

③rye bread

④doughnut

⑤sandwich

⑥farmer bread

⑦French bread

⑧Danish pastry

⑨puff pastry

⑩puff

⑪toast bread

⑫hot dog

❸ "比一比"
看谁说得好。
S：Shop Assistant
C：Customer
C wants to buy bread. S is helping C.

❹ "说一说"（用英语）
你最喜欢吃哪种面包？为什么？
你最喜欢哪个季节？为什么？
你最喜欢哪种颜色？为什么？

❺ "练一练"
请将下面的中文翻译成英文。
全麦面包，面包片，农夫包，裸麦面包，三明治，松饼，吐司面包，牛角包，热狗。

❻ "找一找"
请用中文列出对话中的词组和短语。

Supplementary Knowledge

A Bakery in Shanghai

拓展阅读

After a great start in 2000 in Shanghai International Trade Centre, we now have a 1,200 m² bakery facility from where we supply bakery goods, chocolate, pastries, cookies, ice cream cups and tartlets in any size, shape or flavor you are looking for. Par-baked frozen, frozen and ready-to-use are available on a daily basis, with experience in large volume production for events such as the 2010 Shanghai Expo., Shanghai Masters Tennis Tournament, F1, Shanghai Open Golf Tournament and many others.

With a team of 40 bakers as well as a German and a Dutch master bakers, we produce a wide range of bakery products from mostly imported ingredients. European seasonal specialties for Christmas, Easter and Chinese New Year can be found in our assortment.

Most of our clients are 4 and 5 star hotels, international restaurants and many supermarket chains in Shanghai, Suzhou, and Wuxi.

Bastiaan Bakery Co. Ltd. is now the largest and oldest European bakery in Shanghai and is the choice supplier of many renowned international hotel chains, supermarket chains and restaurant chains in the city.

We are QS and ISO 22000 licensed for 6 years now.

BASTIAAN BAKERY, YOUR CHOICE FOR HIGH QUALITY BAKERY PRODUCTS.

(*Bakery's Story*, http://www.bastiaanbakery.com.cn)

Project 2
Ordering a Birthday Cake

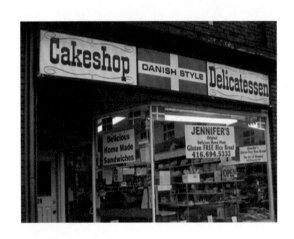

Task 1 任务一：
调查并总结蛋糕店里的生日蛋糕的分类方法，并用英语把它们说出来。（采用小组讨论形式，最后用英语在讲台上表达）

Task 2 任务二：
店员如何向顾客介绍店里的蛋糕产品？假如用英语你怎样说？（采用小组讨论形式，最后用英语在讲台上表达）

Task 3 任务三：
订制生日蛋糕要不要预付订金？如何预付订金？（采用小组讨论形式，最后用英语在讲台上表达）

Task 4 任务四：
店员如何与顾客道别？假如用英语你怎样说？（采用小组讨论形式，最后用英语在讲台上表达）

Which do you like most?

Ordering a Birthday Cake Project ❷ 项目二

birthday cake   Black Forest cake   Angel Food cake

chocolate cake   Tiramisu   mousse cake

cheese cake   Swiss roll   sponge cake

chiffon cake   fruit cake   wedding cake

Dialogue

Mrs. White, together with her daughter, Linda, goes to a bakery to order a birthday cake for Linda.

S: Shop Assistant   W: Mrs. White   L: Linda

S: Can I help you?

W: Yes. I'd like to order a birthday cake.

S: When will the birthday party be held?

W: Next Sunday.

S: What kind of flavor do you like?

W: Chocolate flavor.

S: What's the size of the cake?

Let's listen to the dialogue.

W: 6 pounds is enough. How much does one pound cost?

S: 25 yuan per pound.

L: Mom, look, I like this cake.

W: The one with penguin?

L: Yes, er… er… I also like this one.

W: Which one?

L: The one with bear.

W: Linda, you can choose only one cake.

L: Oh, then, the one with bear.

W: When will the cake be ready for me to pick up?

S: At 2:00 o'clock in the afternoon next Sunday. Is that all right?

W: That will be fine. Do you require a deposit to confirm a reservation?

S: Yes. Please pay 100 yuan now, and the rest next Sunday.

W: OK. Here you are.

S: Here is the receipt. Please keep it well and bring it back next Sunday.

W: Yes, I will.

S: Would you please write down your name and telephone number here?

W: Yes. Of course.

S: Thank you. Goodbye.

W: Goodbye.

Notes

1. When will the birthday party be held?
   生日派对什么时候举行?
2. What kind of flavor do you like?
   您要什么口味的?
3. What's the size of the cake?
   您要多大的蛋糕?
4. Is that all right?
   可以吗?
5. Here you are.
   给您。(把某物递给某人时说的话)

Ordering a Birthday Cake Project 2 项目二

New Words and Expressions

1. cake ［keik］ n. 蛋糕；糕饼
2. flavor ［'fleivə］ n. 风味，滋味
3. size ［saiz］ n. 大小，尺寸
4. chocolate ［'tʃɔkəlit］ n. 巧克力；adj. 用巧克力制的
5. pound ［paund］ n. 英镑（英国的货币单位）；磅（重量单位）
6. enough ［i'nʌf］ adv. 足够地，充足地；adj. 充足的，足够的
7. cost ［kɔst］ vi. 价钱为；花费 n. 价钱；代价；花费
8. per ［pəː, pə］ prep. 每（表示比率）（尤指数量、价格、时间）
9. pay ［pei］ vt. & vi. 付款
10. require ［ri'kwaiə］ vt. & vi. 要求；需要
11. reservation ［ˌrezə'veiʃən］ n. 预订，预约
12. receipt ［ri'siːt］ n. 收据；发票
13. Black Forest cake ［blæk 'fɔrist keik］ 黑森林蛋糕
14. Tiramisu ［ˌtirəmi'suː］ 提拉米苏
15. mousse cake ［muːs keik］ 慕斯蛋糕/慕思蛋糕/木司蛋糕
16. cheese cake ［tʃiːz keik］ 芝士蛋糕，奶酪蛋糕
17. Swiss roll ［swis rəul］ 瑞士蛋糕卷
18. sponge cake ［spʌndʒ keik］ 海绵蛋糕
19. chocolate cake 巧克力蛋糕
20. chiffon cake ［'ʃifɔn keik］ 戚风蛋糕
21. Angel Food cake ［'eindʒəl fuːd keik］ 天使蛋糕
22. fruit cake ［fruːt keik］ 水果蛋糕
23. cake decoration ［keik ˌdekə'reiʃən］ 蛋糕装饰
24. hold a party 举办聚会
25. What kind of... 哪一种……
26. be ready 准备好
27. the rest 剩下的（人或物）；其他的（人或物）
28. pick up 拿起，捡起；取（给）
29. write down 写下；记下

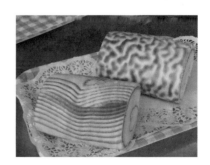

Grammar

1. how much 和 how many 的区别

    how much 表示数量，用来修饰不可数名词，也可单独使用。例如：

    How much money do you get?

    你拿到多少钱？

    How much is the eraser?

    这块橡皮擦多少钱？

    How much?

    多少钱？

    how many 也是表示数量，用来修饰可数名词的复数，它的句式是"How many + 复数名词 + 一般疑问句 + ?"，例如：

    How many books are there on the desk?

    书桌上有多少本书？

    How many days are there in a week?

    每周有几天？

2. pay，cost 和 spend 表示"花费金钱"时的区别

    pay 和 spend 的主语必须是人，cost 的主语是物或某种活动。例如：

    "我花 20 元买了这本书"可译为：

    I spent twenty yuan on this book.

    I paid twenty yuan for this book.

    This book cost me twenty yuan.

    注意：cost 的过去式及过去分词都是 cost，并且不能用于被动句。

Exercises

 ❶ "想一想"

我做过哪种蛋糕？能用英语说出来吗？

 ❷ "试一试"

请找出与中文对应的实物，将序号写在相应的中文后面。

A. 提拉米苏　　　B. 天使蛋糕　　　C. 巧克力蛋糕　　　D. 生日蛋糕

E. 黑森林蛋糕　　F. 婚礼蛋糕　　　G. 戚风蛋糕　　　　H. 奶酪蛋糕

I. 水果蛋糕　　　J. 瑞士蛋糕卷　　K. 木司蛋糕　　　　L. 海绵蛋糕

Ordering a Birthday Cake Project ❷ 项目二

①Tiramisu　②mousse cake　③chocolate cake
④birthday cake　⑤Black Forest cake　⑥Angel Food cake
⑦fruit cake　⑧wedding cake　⑨chiffon cake
⑩cheese cake　⑪Swiss roll　⑫sponge cake

 ❸ "比一比"
看谁说得好。
S：Shop Assistant C：Customer
C wants to buy a cake. S
is helping C.

 ❹ "说一说"（用英语）
你们看过 NBA 吗？
最喜欢哪个球星呢？

15

❺ "练　练"

请将下面的对话翻译成英语。

母亲：你喜欢哪种蛋糕？
儿子：我喜欢科比的，还有詹姆斯的。
母亲：你只能选其中一种。
儿子：哦，那就詹姆斯的那种吧。

❻ "找一找"

请用中文列出对话中的词组和短语。

Supplementary Knowledge

Celebration Cakes

When it comes to a celebration such as a birthday or graduation, the cake is always a key point of interest. A birthday party without a delicious birthday cake just wouldn't be the same.

You can never be too old for a cake, so why not go that one step further and get a fantastic customized one from us here at The Cake Shop? With 25 years' experience in the industry we can design and create whatever style of cake you are after. Your idea will literally be brought to life.

We've created many celebration cakes in a range of shapes, sizes and designs. So whether it's children's birthday cakes, graduation cakes, christening cakes or women's cakes that you are after, we can cater for your need.

We offer delivery service to most areas in the Midlands including Oxfordshire, Warwickshire, Buckinghamshire, Wiltshire, Gloucestershire and Berkshire. Alternatively you can choose to collect your wedding cake from either of our shops.

An extremely distinctive event deserves a unique cake, so have a look online today and make sure your next event includes a custom-made celebration cake from The Cake Shop!

(Celebration Cakes, The Cake Shop, http：//www.the-cakeshop.co.uk/newshop/celebration-cakes.asp)

Project 3
Visiting a Bread Factory

Task 1 任务一：
在面包厂里，面包师应该如何招呼进厂参观的客人？用英语怎样说？（采用小组讨论形式，最后用英语在讲台上表达）

Task 2 任务二：
面包师如何向参观者介绍工厂里面包的制作过程？假如用英语你怎样说？（采用小组讨论形式，最后用英语在讲台上表达）

Task 3 任务三：
用英语写出你知道的制作面包的设备名称。（采用小组讨论形式，最后写在黑板上，写得最多的为本节课的优胜组）

Task 4 任务四：
看到香喷喷的面包产品后，如何描述和称赞产品？假如用英语你怎样说？（采用小组讨论形式，最后用英语在讲台上表达）

Which do you like most?

a bread factory

mixer

bowl

dough

divider

rounder

molder

pan

final proofer

oven

products

slider

Dialogue

A group of tourists are visiting a bread factory. Jack serves as their guide. Factory Director Richard is showing them around.

R: Richard **J**: Jack **T1**: tourist 1
T2: tourist 2

R: Welcome to our factory. I'm Richard.
J: Hello, Richard. My name is Jack.
R: Nice to meet you, Jack.
T1: Hi, Richard. When was the factory built?
R: In 1992.

Let's listen to the dialogue.

Visiting a Bread Factory Project 3 项目三

T1: Oh. More than 20 years.

R: Yes, it's a long time.

T2: How many years have you worked in this factory, Richard?

R: I've worked here since 1998.

T2: Really? Then you are an experienced baker.

R: You can say that again! This way, please. Look, we weigh ingredients here and then place all dry ingredients into a mixer or a bowl, add water and mix the dough until they are well developed.

J: And then?

R: Look, the dough is put there to ferment.

T1: What's this, Richard?

R: That's a molder. After the dough has been fermented, it is loaded into the divider to cut into pre-determined weights. Then knock back and shape the dough. Well, this is the final proofer and the dough is proofing here. After that, the dough is ready for baking. The last step is to cool and package the bread over there.

T1: The bread-making process sounds really complicated. How long will the whole process last?

R: About 5 hours.

T1: Oh, I see.

R: Look. These are the products.

T1: Wow, how attractive they are! And they smell good. Thank you for your introduction.

R: You are welcome. Goodbye.

J: Goodbye.

Notes

1. Welcome to our factory.
   欢迎来到我们工厂参观。

2. You are an experienced baker.
   你是一位经验丰富的面包师。

3. You can say that again!
   说得没错。(表示赞同对方说的话，而非字面上的意思：你可以再说一遍！)

4. How long will the whole process last?
   整个面包制作过程要多久？

5. And they smell good.
   而且它们闻起来很香。

New Words and Expressions

1. factory ['fæktəri] n. 工厂；复数：factories
2. baker ['beikə] n. 面包师；烤炉
3. weigh [wei] vt. 称……的重量
4. mixer ['miksə] n. 搅拌机；混合器
5. bowl [bəul] n. 搅拌缸
6. ingredient [in'gri:diənt] n. 原料，（混合物的）组成部分
7. dough [dəu] n. 生面团
8. shape [ʃeip] n. 形状；vi. 使成形
9. step [step] n. 步；步骤
10. divide [di'vaid] vt. & vi. 分割，切块
11. proof [pru:f] vi. 醒发，最后发酵
12. package ['pækidʒ] vt. 把……包成一包
13. product ['prɔdʌkt] n. 产品
14. attractive [ə'træktiv] adj. 有吸引力的
15. smell [smel] vt. & vi. & link-v. 嗅，闻；闻出
16. ferment [fə'ment] vt. & vi. 使……起发，发酵
17. oven ['ʌvən] n. 烤炉
18. rounder ['raundə] n. 面团滚圆机
19. molder ['məuldə] n. 面团成型机
20. introduction [,intrə'dʌkʃən] n. 介绍
21. bread-making process [bred 'meikiŋ 'prəuses] 面包制作过程
22. an experienced baker 一位经验丰富的面包师

Grammar

1. 现在完成时的"未完成"用法
I've worked here since 1998.
我从1998年起就在这里工作了。

　　这句话使用的是现在完成时持续性的用法（肯定句、疑问句中的谓语动词必须是延续性动词）：表示过去已经开始，持续到现在的动作或状态。此时常与"for + 一段时间"或"since + 过去的时间点"等时间状语连用。

　　肯定句谓语动词的构成："have/has + 过去分词"，第三人称单数作主语时用 has，其余的用 have。例如：
I have known him for ten years.
我已经认识他10年了。
He has known her for ten years.
他已经认识她10年了。

Visiting a Bread Factory Project 3 项目三

2. 感叹句

How attractive they are!
它们多么有吸引力啊!

这是一个感叹句。感叹句通常由 what 或 how 引导，表示赞美、惊叹、喜悦等感情。

what 修饰名词，how 修饰形容词、副词。感叹句的结构主要有以下几种：

1) How + 形容词 + 主语 + 谓语，例如:
   How beautiful the girl is!
   这女孩多美啊!

2) What + a/an + 形容词 + 可数名词单数 + 主语 + 谓语，例如:
   What a beautiful girl she is!
   这女孩多美啊!

3) What + 形容词 + 不可数名词或可数名词复数 + 主语 + 谓语，例如:
   What beautiful flowers they are!
   这些花多美啊!
   What cold water it is!
   这水好冷啊!

Exercises

① "想一想"
我都用过哪些设备了? 能用英语说出来吗?

② "试一试"
请找出与中文对应的实物，将序号与在相应的中文后面。

A. 搅拌机   B. 面团分割机   C. 成型机   D. 烤炉
E. 面包切片机   F. 醒发箱   G. 搅拌缸   H. 面包厂
I. 产品   J. 面团   K. 面包烤盘   L. 面团滚圆机

①slider

②products

③a bread factory

④oven

⑤dough

⑥pan

⑦rounder

⑧molder

⑨mixer

⑩final proofer

⑪divider

⑫bowl

❸ "比一比"
看谁说得好。
E：A Bread Factory Engineer
V：Visitor
V is visiting a bread factory now.
E is showing him around.

❹ "说一说"（用英语）
我学习英语3年了。
我从2012年起住在广州。
多美丽的城市啊！
多健康的一个孩子啊！

❺ "练一练"
请将下面的中文翻译成英文。
醒发箱
超过20年
面包制作过程
一个经验丰富的面包师

❻ "找一找"
请用中文列出对话中的词组和短语。

Supplementary Knowledge

Sandwich

At lunch time, and sometimes at other times in the day, people in the UK often eat a "sandwich". This consists of two pieces of bread and a filling. The bread is usually buttered or spread with mayonnaise, and the filling is usually meat or cheese, often served with lettuce.

However, there are literally hundreds of different types of sandwiches, and each variation has its own flavour. Some of the most popular and famous sandwiches in the UK are: BLT (Bacon, Lettuce and Tomato, usually served with mayonnaise), Ploughman's (originating from a tradition amongst farm-workers, and containing Cheddar cheese, pickle and salad), Tuna and Egg sandwiches.

Sandwiches of all varieties are extremely popular, and quick and easy to eat. In fact, British people eat 2.8 billion each year—not bad for a population of only 60 million people! Today everyone eats sandwiches, but it was not always like that. Amazingly, the humble sandwich that we know today started life as a snack for England's super-rich! The sandwich has a very interesting and humorous history!

In 1762 the first written record of the word "sandwich" appeared in the diary of English author Edward Gibbons, who remembered seeing the wealthiest elite in the country eating "a bit of cold meat" between pieces of bread. Gibbons did not think this was very appropriate behaviour for such men!

The snack was named after the Fourth Earl of Sandwich (an Earl was a wealthy aristocrat, who generally owned a lot of land and had political power). Sandwich was a frequent gambler, and was so addicted to gambling that he would often refuse to stop even to eat meals! To avoid having to stop gambling, the Earl of Sandwich asked the cooks at his gambling club to prepare him a meal consisting of beef between two slices of bread, so that he always had one hand free to play cards and gamble, and his hands wouldn't become dirty from the meat.

When other men saw what he was eating, they began to order "the same as Sandwich", and so the sandwich was born—beginning as a snack for some of the wealthiest men in England! It quickly became popular and widespread as a quick and easy food to eat.

（英国总领事馆文化教育处资料）

Project 4
Visiting a Cake Shop

Task 1 任务一：
在蛋糕店里，店员应该如何招呼进店的参观者？用英语怎样说？（采用小组讨论形式，最后用英语在讲台上表达）

Task 2 任务二：
店员在带领参观者参观的过程中常常会说"请走这边"。假如用英语你怎样说？（采用小组讨论形式，最后用英语在讲台上表达）

Task 3 任务三：
市面上有各种各样的蛋糕产品，但它们实际上可以归纳为三类，你知道是哪三类吗？用英语你怎样说？能写出来吗？（采用小组讨论形式，最后用英语在讲台上说出来并写在黑板上，发音和书写都正确的为本节课的优胜组）

Task 4 任务四：
参观完毕后，如何向对方表达谢意？假如用英语你怎样说？（采用小组讨论形式，最后用英语在讲台上表达）

Which do you like most?

Visiting a Cake Shop Project 4 项目四

cake chocolate cake butter cake

wedding cake sponge cake Angel Food cake

cheese cake fruit cake foam cake

chiffon cake birthday cake ⑫mousse cake

Dialogue

Lucy is visiting a cake shop now.

S: **Shop Assistant**   L: **Lucy**

S: Welcome to our shop. I'm Nancy.

L: Hello, Nancy. My name is Lucy.

S: Nice to meet you, Lucy. This way, please.

L: How many types of cakes are there in the world?

S: There are nearly hundreds of different varieties of cakes today, but they all belong to three main types: ① butter cakes, ② foam cakes (Angel Food and sponge), and ③ chiffon cakes.

Let's listen to the dialogue.

L: What types of cakes do you have?

S: We have wedding cakes, birthday cakes, chocolate cakes and so on.

L: What shapes and sizes do you make?

S: We make cakes in all shapes and sizes. Besides, cakes can be decorated with company logos, images and messages.

L: How much would you charge for your cakes?

S: It's hard to say. The price mainly depends on the quality.

L: Thank you for your introduction.

S: My pleasure. Come again when you need to buy cakes. I'll give you a discount.

L: Sure. Thank you so much. Goodbye.

S: Goodbye.

Notes

1. This way, please.
   这边请。
2. There are nearly hundreds of different varieties of cakes today.
   现在有近百个不同品种的蛋糕。
3. What types of cakes do you have?
   你们有什么类型的蛋糕?
4. How much would you charge for your cakes?
   你们的蛋糕卖多少钱?
5. The price mainly depends on the quality.
   价格主要取决于质量。
6. I'll give you a discount.
   我会给你打个折扣。

我们来看看下面的注释。

New Words and Expressions

1. type [taip] *n.* 类型
2. foam [fəum] *n.* 泡沫
3. sponge [spʌndʒ] *n.* 面种；海绵；海绵状物
4. butter [ˈbʌtə] *n.* 牛油（天然牛油）
5. batter [ˈbætə] *n.* 面糊（用鸡蛋、牛奶、面粉等调成的糊状物）
6. charge [tʃɑːdʒ] *vi.* 索价；收费
7. price [prais] *n.* 价格，价钱
8. mainly [ˈmeinli] *adv.* 大部分地；主要地
9. quality [ˈkwɔliti] *n.* 质量，品质。复数：qualities
10. hundreds of 好几百
11. foam cake 乳沫类蛋糕
12. belong to 属于；归于
13. butter cake 奶油蛋糕
14. such as （表示举例）例如，诸如此类的，像……那样的（相当于 like 或 for example）

Grammar

There be 句型的用法

1. 构成

There be 句型表示"某处有（存在）某人或某物"，其结构为"There be（is，are，was，were）+ 名词 + 地点状语"。例如：

There are fifty-two students in our class.

我们班有 52 个学生。

There is a pencil in my pencil-box.

我的铅笔盒里有一支铅笔。

There was an old house by the river five years ago.

5 年前那条河边有一幢旧房子。

2. 否定句式

There be 句型否定句式的构成和含有 be 动词的其他句型一样，在 be 后加上"not"；也可用"no"来表示。例如：

(1) There is an orange in her bag.

她的包里有一只橘子。

There isn't an orange in her bag.

There is no orange in her bag.

（2）There are some oranges in her bag.

她的包里有一些橘子。

There aren't any oranges in her bag.

There are no oranges in her bag.

（3）There is some juice in the bottle.

瓶子里有一些果汁。

There isn't any juice in the bottle.

There is no juice in the bottle.

3．注意事项

（1）There be 句型中，be 动词的形式要和其后的主语在人称和数上保持一致。如果句子的主语是单数可数名词，或是不可数名词，be 动词用"is"或"was"。例如：

There is a basketball in the box.

There is a little milk in the glass.

如果句子的主语是复数名词，be 动词就用"are"或"were"。例如：

There are three birds in the tree.

There were many people in the street yesterday.

如果有两个或两个以上的名词作主语，be 动词要和最靠近它的那个主语在数上保持一致，这就是我们常说的"就近原则"。例如：

There is an orange and some bananas in the basket.

There are some bananas and an orange in the basket.

（2）There be 句型和 have/has 的区别：

There be 句型表示"存在"；have/has 表示"拥有"、"所有"。例如：

桌子上有三本书。

There are three books on the desk.

我有三本书。

I have three books.

Exercises

1 "想一想"
哪种蛋糕最好吃？能用英语说出来吗？

2 "试一试"
请找出与中文对应的实物，将序号写在相应的中文后面。

A. 巧克力蛋糕　　B. 水果蛋糕　　C. 天使蛋糕　　D. 婚礼蛋糕
E. 海绵蛋糕　　　F. 生日蛋糕　　G. 戚风蛋糕　　H. 奶油蛋糕
I. 芝士/奶酪蛋糕　J. 慕斯蛋糕　　K. 乳沫蛋糕　　L. 蛋糕

①Angel Food cake

②cheese cake

③fruit cake

④foam cake

⑤chiffon cake

⑥birthday cake

⑦cake

⑧chocolate cake

⑨butter cake

⑩wedding cake

⑪sponge cake

⑫mousse cake

❸ "比一比"
看谁说得好。
S：Shop Assistant
C：Customer
C is visiting a cake shop. S is showing him around.

❹ "说一说"（用英语）
教室里有30个学生。
书桌上有5本书。
房间里有一张床。

❺ "议一议"
英文中的面包、蛋糕是论个的还是论块的？

Key Phrases：
a piece of bread, two pieces of bread
a loaf of bread, two loaves of bread
a piece of cake, a cake

❻ "找一找"
请用中文列出对话中的词组和短语。

Supplementary Knowledge

Do You Know Anything about Bread?

Bread makes an appearance on just about every Western menu. There are so many different kinds of bread that sometimes just choosing the bread to accompany your main meal can be an arduous task. If you are overseas, it is useful to know the different kinds of bread and the proper ways to eat them.

拓展阅读

Basically, there are 3 kinds of bread：

1. Yeast Bread

Yeast bread is recognizable by the tiny air pockets all through them. The yeast in the dough causes the bread to rise during baking—giving it that "loaf" look. Pan bread such as raisin bread and whole-wheat bread are considered to be yeast bread. Pizza base and hamburger bun are also yeast bread.

2. Quick Bread

Quick bread takes less time to prepare than yeast bread. This bread also rises during baking, but it is caused by baking powder instead of yeast. Examples of quick bread are corn bread,

doughnuts, muffins and pancakes.

3. Flat Bread

This kind of bread is appropriately named as its appearance is indeed flat. Flat breads include tortillas (Mexican), chapatti (Indian) and pita bread (Middle Eastern). They usually have an empty center in them for fillings and sauces.

The proper way to eat bread:

In a proper western restaurant, the bread plate is always on your left with your butter knife on it. When eating bread, you should break the bread into small sizes. It is rude to cut your bread or bite from it. You should spread butter only on the piece of bread you are about to eat (not on all the bread on your plate). It is OK if breadcrumb drops on your plate or on the table. The waiter will clear them away for you.

(Jean,《视听英语 Ladder AI》,http://www.sina.com.cn,2004/07/28)

Module 2: Formulas and Making

模块二

配方与制作

You are going to:

1. learn how to make different kinds of bread and cakes.
2. know many different formulas of bread and cakes.
3. get more knowledge about the ingrodionto for balking.

要动脑筋哦!

Project 5
Bread's Formula

Task 1 任务一：
认识配方表中各种原料的名称，并用英语表达出数量。（采用小组讨论形式，最后用英语回答教师的问题）

Task 2 任务二：
看得明白制作的流程及注意事项，并掌握关键步骤的英语表达。（采用小组讨论形式，最后用英语在讲台上表达）

Task 3 任务三：
用英语写出本产品所需要的原料名称及数量。（采用小组讨论形式，最后用英语在讲台上表达，知道最多的为本节课的优胜组）

Task 4 任务四：
大家讨论一下，我们在实训室里是怎样做面包的呢？（采用小组讨论形式，最后用英语在讲台上表达）

Which do you like most?

Text 1

大家看看，下面的这个配方是不是很熟悉？没错，它就是我们在面包实验中所做的无盖吐司方包（甜）的配方。

面包实验：无盖吐司方包（甜）				
姓名：		实验日期：		
面种	原料	%	克	制作过程
	高筋粉	70	700	1. 要求面种温度26℃，求解适用水温。 2. 面种搅拌至卷起阶段。记录面种温度：_____℃。 3. 面种于28℃、75%相对湿度下，发酵约2.5小时。 4. 记录面种发酵时间：____；发酵后面种温度：____℃。 5. 记录发酵后面种状态：_____。
	低筋粉			^
	水	65	455	^
	酵母	0.8	8	^
	盐			^
	适用水温 = 理想面团温度____×2 –（室温____+ 粉温____+ 摩擦升温____）= ____。			
主面团	原料	%	克	制作过程
	高筋粉	30	300	1. 要求主面团温度28℃，求解适用水温。 2. 主面团搅拌加料顺序：发酵后面种 + 水 + 糖 + 蛋 + 奶粉 + 改良剂—慢速拌匀—加入面粉、酵母—慢速拌匀—中速搅拌至面筋扩展—加入奶油—中速搅拌至面筋完全扩展—加入盐，搅拌2分钟至均匀。 3. 主面团在室温下持续发酵30分钟。 4. 主面团分割，分割重量为400克/个×4个。 5. 面团滚圆，台面松弛30分钟。 6. 面团擀薄排气，卷起成型后入模。 7. 最后醒发：36℃，85%湿度，80～90分钟。 8. 烘烤：上火160℃，下火190℃，35～40分钟。 9. 出炉后趁热脱模，立于冷却架上冷却至室温。
	低筋粉			^
	水	55	95	^
	酵母	0.2	2	^
	盐	1.5	15	^
	糖	20	200	^
	奶油	6	60	^
	鸡蛋	5	50	^
	奶粉	4	40	^
	改良剂	0.3	3	^
	总量	192.8	1 928	^
	适用水温 = 理想面团温度____×4 –（室温____+ 粉温____+ 摩擦升温____+ 发酵后面种温度____）= ____。			
搅拌后主面团温度：				
主面团搅拌至完全扩展后面团状态：				
最后醒发：温度____℃，相对湿度____%，醒发用时____分钟。				
最后醒发完成后面团状态：				
烘烤：上火____℃，下火____℃，烘烤时间____分钟。				
实验结果分析：				

任务一：

请试着把上述配方中的材料翻译成英文。（不懂的单词或词组可查阅本课的 "New Words and Expressions"）

分析各组完成任务一的情况。

任务二：

阅读下面无盖主食面包（咸）的配方及制作过程，并试着译成中文。

A Formula of White Pan Bread

	Ingredients	Baker's %	Grams	Mixing procedure
Sponge	Bread flour	80	1,600	Dissolve yeast in part of water, mix ingredients to have a sponge, then ferment for 2~3 hours.
	Water	65	1,040	
	Instant yeast	1.0	20	
	Emulsifier	0.3	6	
Dough	Bread flour	20	400	Mix ingredients into the sponge until the gluten formed.
	Water	60~62	160~200	
	Sugar	8	160	
	Milk powder (nonfat)	4	80	
	Shortening	5	100	Add shortening and mix the dough for the gluten's being developed.
	Salt	1.5	30	Add salt and mix the dough smooth.

Procedures to Follow

1. Give the dough a rest period of 20 minutes.
2. Divide and scale the dough into units. Weight depends on the size of loaf pans. Round the dough pieces and allow them to rest for 15 minutes.
3. Sheet the dough into pieces. Starting from top portion roll the dough piece; the length of the rolled dough must be the same as that of the loaf pan. Seams should be sealed properly. Another method is to fold one end of the dough to the other end; roll down and seal edges tightly; make sure the seam is closed at the bottom end.
4. Molded unit of the dough is then placed in a greased loaf pan.
5. Give the dough enough proofing. This can be done by the use of proof room.
6. Bake at 410°F ~ 420°F. The baked bread is removed from the pan immediately to avoid sweating.

　　White loaf could also be made in either round or loaf form. Shaped and molded dough are given 3/4 proof and sliced through top center. The round shaped may be cut across the center or

across the side. After the top is worked with water and dusted lightly with cake flour, it is baked as regular white pan bread.

New Words and Expressions

1. baker's % 烘焙百分比（专门用于烘焙生产中）
2. gram [græm] n. （重量单位）克
3. procedure [prəˈsiːdʒə(r)] n. 程序，手续；工序，过程，步骤
4. bread flour [bred ˈflauə] 高筋面粉；面包专用面粉
5. yeast [jiːst] n. 酵母（菌）；酵母粉，酵母饼，酵母片
6. dissolve [diˈzɔlv] vt. 使溶解；使（固态物）溶解为液体
7. gluten [ˈgluːtn] n. 面筋；麸质
8. develop [diˈveləp] vt. 使发展；使发育；开发；培育
9. sugar [ˈʃugə(r)] n. 糖；一块（茶匙等）糖
10. shortening [ˈʃɔːtniŋ] n. 酥油；雪白奶油
11. milk powder 奶粉
12. nonfat [ˈnʌnˈfæt] 脱脂的
13. optional [ˈɔpʃənl] adj. 可选择的；随意的，任意的
14. period [ˈpiəriəd] n. 时期；（一段）时间；学时
15. scale [skeil] vt. 测量；称量
16. unit [ˈjuːnit] n. 单位，单元
17. weight [weit] n. 重量，体重
18. depend on 取决于
19. loaf pan 面包听
20. round [raund] vt. & vi. 使成圆形
21. sheet [ʃiːt] vi. 成片展开，擀面，擀薄
22. portion [ˈpɔːʃən] n. 一部分
23. seam [siːm] n. 接缝，接合处；线缝；裂缝
24. molded [ˈməuldid] adj. 成形的
25. grease [griːs] vt. 涂油脂于，用油脂润滑
26. proof room 醒发室
27. place [pleis] vt. 放置
28. fold [fəuld] vt. 折叠；合拢

Learning Tips

英语和美语在读音上的差异：

1. 在 ask, can't, dance, fast, half, path 这一类单词中，英国人将字母 a 读作 [ɑː]，而美国人则读作 [æ]。所以这些词美国人读起来就成了 [æsk]、[kænt]、[dæns]、[fæst]、[hæf] 和 [pæθ]。

❶ "做一做"
请用中文列出该配方中所需的原料及数量。

❷ "找一找"
在配方表中找出意思相近的英文内容。

把酵母溶在水里：_____
发酵2~3小时：_____
整形面团：_____
一个涂了油的面包听：_____

❸ "议一议"
大家讨论一下，我们在实训室里是怎样做面包的呢？

❹ "试一试"
请用中文简单描述该面包的制作程序。

Text 2

A Formula of Whole Wheat Bread

	Ingredients	Baker's %	Grams	Mixing procedure
I	Whole wheat flour	40	800	Suspend yeast in part of the water. Place all ingredients of part I in the bowl and mix until partially developed.
I	Bread flour	60	1,200	Suspend yeast in part of the water. Place all ingredients of part I in the bowl and mix until partially developed.
I	Yeast (dry)	1	20	Suspend yeast in part of the water. Place all ingredients of part I in the bowl and mix until partially developed.
I	Sugar (brown)	10	200	Suspend yeast in part of the water. Place all ingredients of part I in the bowl and mix until partially developed.
I	Water	63	1,260	Suspend yeast in part of the water. Place all ingredients of part I in the bowl and mix until partially developed.
I	Molasses	1	20	Suspend yeast in part of the water. Place all ingredients of part I in the bowl and mix until partially developed.
II	Shortening	4	80	Add shortening and mix the dough for the gluten's being developed.
III	Salt	2	40	Add salt and mix the dough smooth, then ferment for 1.5~2 hours.

Procedures to Follow

1. Divide and scale the dough into units. Weight depends on the size of loaf pans, round the dough pieces and allow them to rest for 15 minutes.
2. Sheet the dough into pieces. Starting from top portion roll the dough piece; the length of the rolled dough must be the same as that of the loaf pan; seams should be sealed properly. Another method is to fold one end of the dough to the other end; roll down and seal edges tightly; seal the seam properly.
3. Molded unit of the dough is then placed in a greased loaf pan.
4. Give the dough enough proofing. This can be done by the use of steam plate.
5. Bake at 410 ℉ ~ 420 ℉. The baked bread is removed from the pan immediately to avoid sweating.

New Words and Expressions

1. suspend ［sə'spend］ vt. 悬浮，溶解
2. brown sugar 红糖
3. partially ［'pɑːʃəli］ adv. 部分地
4. add ［æd］ vt. 增加；补充
5. smooth ［smuːð］ adj. 光滑的；流畅的；柔软的
6. punch ［pʌntʃ］ vt. 用拳猛击；翻动面团或翻面
7. molasses ［mə'læsiz］ n. 糖浆；糖蜜

Grammar

被动语态

被动语态由"助动词 be + 及物动词的过去分词"构成。人称、数和时态的变化是通过 be 的变化表现出来的。现以 teach 为例说明被动语态在各种时态中的构成。

一般现在时：am/is/are + taught
一般过去时：was/were + taught
一般将来时：will/shall be + taught
现在进行时：am/is/are being + taught
过去进行时：was/were + being + taught
现在完成时：have/has been + taught

歌诀是：被动语态 be 字变，过去分词跟后面。
被动语态表示主语是动作的承受者，即行为动作的对象。巧记为：被动、被动、主被动。

例 1：
   主动语态：人们说英语。People speak English in many countries.
   被动语态：英语被说。English is spoken in many countries.
例 2：
   主动语态：我们造这座桥。We built this bridge last year.
   被动语态：这座桥被建造。This bridge was built last year.

Choose the best answer.

1. Chinese _____ by the largest number of people.
   A. speak　　　B. is speaking　　　C. speaks　　　D. is spoken
2. The boy _____ to get supper ready after school.
   A. were told　　B. is telling　　　C. was told　　　D. tells

3. This dictionary mustn't _____ from the library.
   A. take away　　　B. taken away　　　C. are taken away　　D. be taken away
4. —Your coat looks nice. Is it _____ cotton?
   —Yes. It's _____ Shanghai.
   A. made of; made by　　　　　B. made of; made in
   C. made for; made by　　　　　D. made for; made in
5. The radio says a wild animal zoo is to _____ in our city.
   A. be building　　B. build　　C. be built　　D. built

Exercises

❶ "做一做"
请用中文列出该配方中所需的原料及数量。

❷ "找一找"
在配方表中找出意思相近的英文内容。

部分发酵：_____
切分面团成单个：_____
击打面团：_____

❸ "试一试"
　　大家一起讨论该面包的制作过程并记录下来：

Supplementary Knowledge

What Is Bread?

Bread in this country has, to everybody's benefit, reached a high standard of purity and hygiene.

Bread is perhaps the most important item in our diet; it has often been called "the staff of life". To give you an idea of the benefit we get from flour and bread, a government survey showed that flour and bread provided us with more energy value, more protein, more iron, more nicotinic acid and more vitamin B1 than any other basic food. Bread comes to us in many interesting shapes and flavours. From the time-honoured "cottage" loaf, to some of the delicious Vienna rolls. Nowadays, the sliced and wrapped loaf is the most popular loaf of all. It is ideal for making sandwiches for picnics, and for workers' lunches. There is, however, an important drawback. If you like your bread with a beautiful rich golden crust on it, do not buy the ready-wrapped variety. One of the nicest things in life is to come home hungry from school or work, and have set before one the fresh, buttered crust from a well-done cottage or coburg loaf.

Bread is such an important part of our lives that it ought to be taken more seriously, and enjoyed to the full. In your town, there are probably a number of bakers. Find the one whose bread you usually enjoy. Besides the ordinary white, wholemeal and wheatmeal loaves, many other kinds are on sale which the baker calls "fancies". There are the "malt" bread, bread with currants, milk loaves (containing milk powder), and various tea bread. Then there is spiced bread, in the form of ginger-bread, but this really comes under the heading of cake, although in Holland it always features on the breakfast table.

Time is marching on in many fields of industry, total mechanisation is the order of the day, and as you have seen, the baking industry is rapidly becoming mechanised. An ordinary loaf needs about three-quarters of an hour in the oven at present. But already, electronic devices are being developed that can bake a loaf, by means of high-frequency heat, in three minutes. A loaf baked so quickly, though, has no time to form a crust—the product is not an attractive one. It would have a great use, though, in international emergencies, such as great earthquakes, floods, etc., when perhaps thousands of people would be in dire need of food. A neighbouring country could make and send huge batches of bread to the stricken area in a very short time. Have you ever thought how much bread you eat in a year? As well as the meat, potatoes, vegetables, etc., you probably eat more than 100kg or nearly twice your own weight.

There is one thing you can be sure of—bread is one of the finest foods that is possible to get. In fact, it would not be an exaggeration to say that we cannot do without it. There are many items of food and luxury, such as ice-cream or sweets that we could well do without, and be far healthier for it. A balanced diet to keep you strong and well in mind and body must always contain that staff of life—good bread.

(*The Story Behind a Loaf of Bread*, http://www.botham.co.uk/bread/bread1.htm)

Project 6
Cake's Formula

Task 1 任务一：
认识配方表中各种原料的名称，并用英语表达出数量。（采用小组讨论形式，最后用英语回答教师的问题）

Task 2 任务二：
看得明白制作的流程及注意事项，并掌握关键步骤的英语表达。（采用小组讨论形式，最后用英语在讲台上表达）

Task 3 任务三：
用英语写出本产品所需要的原料名称及数量。（采用小组讨论形式，最后用英语在讲台上表达，知道最多的为本节课的优胜组）

Task 4 任务四：
大家讨论一下，我们在实训室里是怎样做蛋糕的呢？（采用小组讨论形式，最后用英语在讲台上表达）

Which do you like most?

Text 1

A Formula of Butter Cake

Ingredients	Baker's %	Grams	Mixing procedure
Sugar	100	1,000	Place into mixing bowl and mix in 2nd speed until partially creamed. Scrape down several times in this stage.
Shortening (emulsified)	80	800	
Salt	2	20	
Eggs	88	880	Add eggs for several times and scrape after each addition.
Milk	20	200	Add milk, and mix.
Cake flour	100	1,000	Sift together then add to the bowl and mix to smooth in first speed.
Baking powder	0.6	6	
Flavor	to suit		Add flavor, and mix 3 mins in first speed.

Procedure to Follow

Deposit into loaf pans, grease and line and bake at 350 °F for 30~45 minutes.

New Words and Expressions

1. speed [spi:d] *n.* 速度；变速器，排挡
2. scrape [skreip] *vt.* 擦，刮；擦去
3. stage ['steidʒ] *n.* 阶段
4. emulsified [i'mʌlsifaid] *adj.* 乳化的
5. baking powder ['beikiŋ 'paudə] *n.* 发酵粉，发粉
6. addition [ə'diʃn] *n.* 加，增加
7. sift [sift] *vt.* 筛分；精选
8. deposit [di'pɔzit] *vt. & vi.* 放置，安置

Grammar

一般现在时的用法

　　表示经常的或习惯性的动作，动词用一般现在时。例如：
1. I'm a student. 我是一名学生。
2. We are Chinese. 我们是中国人。
3. We go to work every day. 我们每天去上班。
4. They often play basketball after school. 他们经常放学后打篮球。

Fill in the blanks with the proper form of the words in the brackets.

Example: — What does he do?
　　　　— He is (be) a bus driver.

1. He _____ (be) a waiter.
   He _____ (like) his job.
2. She _____ (be) a nurse.
   She _____ (work) in a hospital.
3. Susan _____ (be) a secretary.
   She _____ (find) information on the Internet.
4. Tom and Mary _____ (be) tour guides.
   They _____ (show) people around.

Exercises

❶ "做一做"
请用中文列出该配方中所需的原料及数量。

刮缸: _____
用第三挡: _____
发酵粉: _____
搅拌三分钟: _____

❷ "找一找"
在配方表中找出意思相近的英文内容。

❸ "试一试"
请用中文简单描述该蛋糕的制作程序:

Text 2

Swiss Roll

Ingredients	Baker's %	Grams	Mixing Procedure
Eggs	250	1,000	Beat until stiff.
Sugar	100	400	
Cake flour	100	400	Sift cake flour and add them into the beaten eggs/ sugar mixture. Mixing slowly.
Milk	10	40	
Oil	25	100	Add and mix slowly.

Procedures to Follow

1. Pour into a paper-lined cookie sheet (33cm×43cm).
2. Bake at 35°F for 15 minutes. Then cool down.
3. Spread the jelly or cream fillings and roll tightly like a roll.

New Words and Expressions

1. stiff [stif] adj. 发泡的；硬的
2. filling [ˈfiliŋ] n. 填充物；(糕点内的) 馅
3. corn starch [kɔːn staːtʃ] n. 玉米淀粉
4. paper-lined cookie sheet 烘焙纸
5. cool [kuːl] vt. 冷却
6. jelly roll 果酱卷
7. cream filling 打发奶油，果酱，奶油馅
8. powdered sugar 糖粉
9. double boiler [dʌbl bɔilə] n. (美) 双层蒸锅
10. cream [kriːm] vt. 把……搅成糊状（或奶油状）混合物
11. light [lait] adj. 光亮的
12. apply [əpˈlai] vt. 应用，涂；敷

Grammar

Until 的用法

until 前的主句中的动词用延续性动词，后面接时间点。如：I do my homework until 11 o'clock. I slept until my mother called me.

区别于：not...until 直到……才……，Not...until 中主句的动词用短暂性动词，例如：

I won't believe it until I see it with my own eyes.

I didn't go to bed until my mother came back.

I. 翻译下列句子。

1. 他每天都做完作业才睡觉。
_____

2. 直到下午五点半她才见到她姑姑。
_____

3. 他将一直等到明天。
_____

4. 三点钟我才回来。
_____

II. 将下列句子改为 not...till/until 句型。

1. I can do the cooking when you buy something.
_____

2. He will go home after the school is over.
_____

Learning Tips

英语和美语在读音上的差异

2. 在 box, crop, hot, ironic, polish, spot 这一类单词中，英国人将字母 o 读作 [ɔ]，而美国人则将 o 读作近似 [ɑː] 音的 [a]。所以这些词在美国人读起来就成了 [baks]、[krap]、[hat]、[aiˈranik]、[paliʃ] 和 [spat]。

我们来做做下面的练习。

❶ "做一做"
请用中文列出该配方中所需的原料及数量。

打到发泡:＿＿＿＿＿
涂果酱:＿＿＿＿＿
倒入烘焙纸:＿＿＿＿＿
边煮边搅拌:＿＿＿＿＿

❷ "找一找"
在配方表中找出意思相近的英文内容。

❸ "试一试"
请用中文简单描述该蛋糕的制作程序:

Supplementary Knowledge

About Cakes

Cake is a form of bread or bread-like food. In its modern forms, it is typically a sweet baked dessert. In its oldest forms, cake was normally fried bread or cheese cake, and normally had a disk shape. Determining whether a given food should be classified as bread, cake or pastry can be difficult.

Modern cakes, especially layer cakes, normally contain a combination of flour, sugar, eggs, and butter or oil, with some varieties also requiring liquid (typically milk or water) and leavening agents (such as yeast or baking powder). Flavorful ingredients like fruit purées, nuts, dried or candied fruit, or extracts are often added, and numerous substitutions for the primary ingredients are possible. Cakes are often filled with fruit preserves or dessert sauces (like pastry cream), iced with buttercream or other icings, and decorated with marzipan, piped borders or candied fruit.

Cake is often the dessert of choice for meals at ceremonial occasions, particularly weddings, anniversaries, and birthdays. There are countless cake recipes: some are bread-like, some rich and elaborate, and many are centuries old. Cake making is no longer a complicated procedure, while at one time considerable labor went into cake making (particularly the whisking of egg foams), baking equipment and directions have been simplified so that even the most amateur cook may bake a cake.

(Cake, http://en.wikipedia.org/wiki/Cake)

Project 7
Butter Cookies

Task 1 任务一：
认识配方表中各种原料的名称，并能用英语表达出数量。（采用小组讨论形式，最后用英语回答教师的问题）

Task 2 任务二：
掌握制作的流程及注意事项，并掌握关键步骤的英语表达。（采用小组讨论形式，最后用英语在讲台上表达）

Task 3 任务三：
用英语写出本产品所需要的原料名称及数量。（采用小组讨论形式，最后用英语在讲台上表达，知道最多的为本节课的优胜组）

Task 4 任务四：
大家讨论一下，我们在实训室里是怎样做曲奇的呢？
（采用小组讨论形式，最后用英语在讲台上表达）

Which do you like most?

Butter Cookies Project 7 项目七

A Formula of Butter Cookies

Ingredients	Baker's %	Grams	Mixing procedure
Sugar	45	225	Cream well; scrape bottom and side of bowl from time to time.
Salt	1	5	
Butter or margarine	70	350	
Eggs	25	125	Add eggs, and continue creaming for 2~3 minutes. Scrape side and bottom of the mixing bowl.
Cake flour	90	450	Add and mix smoothly.
Bread flour	10	50	

Procedures to Follow

1. Bag out using desired star tube.
2. Variations could be made.
3. Bake approximately at 370°F only until slightly brown edge appears on cookies.

Note

Never overbake. These butter cookies hold shape and sharp edges if mixed accordingly to the instruction and properly bagged.

New Words and Expressions

1. cookie [ˈkʊki] n. 曲奇饼
2. egg whites 蛋白
3. bag out 裱花
4. desired star tube 星型挤花嘴
5. variation [ˌveəriˈeiʃn] n. 不同形状
6. approximately [əˈprɔksimətli] adv. 大约

7. slightly ［ˈslaitli］ adv. 一点点，轻微
8. edge ［edʒ］ n. 边缘
9. appear ［əˈpiə(r)］ vi. 出现
10. overbake ［ˈəuvəˈbeik］ vt. 烘焙过度
11. properly ［ˈprɔpəli］ adv. 恰当地

Grammar

if 从句的用法

1. if 意为 "如果"，引导条件状语从句，可放在主句后，也可放在主句前，常用逗号隔开。

2. 含 if 引导的条件状语从句的复合句中，主句有下列情况时，从句用一般现在时，表示将来意义：当主句为一般将来时态时，如：If your daughter comes, I will call you up. 当主句是祈使句时，如：Please stay at home if it rains tomorrow. 当主句的谓语含有 can，may，must 等情态动词时，如：If he goes on smoking, it may be very bad for his health. 当主句的谓语是 hope，wish，want 等动词时，如：I want to go there if the rain stops.

3. if 还可以当 "是否" 讲，主要用于宾语从句，相当于 whether，从句的时态根据语境来确定。而 if 引导的条件状语从句中，常用一般现在时表将来。如：

— I'll go to school on foot today.
— If you do, you'll be late for class.

Choose the best answer.

1. If he _____ hard, he will get good grades.
   A. study   B. studies
   C. will study   D. studied

2. I want to know if Mary _____ to the party tomorrow.
   A. go   B. went
   C. will go   D. goes

3. Everyone must dress up. If you _____, they won't let you in.
   A. don't   B. won't
   C. can't   D. mustn't

4. Mary will go to Sanya if she _____ a five-day trip.
   A. have   B. had
   C. will have   D. has

5. If he comes late, _____ he will miss the train.
   A. and   B. so
   C. /   D. or

Exercises

❶ "做一做"
请用中文列出该配方中所需的原料及数量。

❷ "试一试"
请用中文简单描述该曲奇的制作程序：

Supplementary Knowledge

Hamburger

The hamburger, a ground meat patty between two slices of bread, was first created in America in 1900 by Louis Lassen, a Danish immigrant, owner of Louis' Lunch in New Haven, Connecticut. There have been rival claims by Charlie Nagreen, Frank and Charles Menches, Oscar Weber Bilby, and Fletcher David. White Castle traces the origin of the hamburger to Hamburg, Germany with its invention by Otto Kuase. However, it gained national recognition at the 1904 St. Louis World's Fair when the *New York Tribune* namelessly attributed the hamburger as, "the innovation of a food vendor on the pike". No conclusive claim has ever been made to end the dispute over the inventor of the hamburger with a variety of claims and evidence asserted since its creation.

The Library of Congress has officially declared that Louis Lassen of Louis' Lunch, a small lunch wagon in New Haven, Connecticut, sold the first hamburger and steak sandwich in the U. S. in 1900. *New York Magazine* states that, "The dish actually had no name until some rowdy sailors from Hamburg named the meat on a bun after themselves years later", noting also that this claim is subject to dispute. A customer ordered a quick hot meal and Louis was out of steaks. Taking ground beef trimmings, Louis made a patty and grilled it, putting it between two slices of toast. Though some critics like Josh Ozersky, a food editor for *New York Magazine*, claim that this sandwich was not a hamburger because the bread was toasted.

Hamburgers are usually a feature of fast food restaurants. The hamburgers served in major fast food establishments are usually mass-produced in factories and frozen for delivery to the site. These hamburgers are thin and of uniform thickness, differing from the traditional American hamburger prepared in homes and conventional restaurants, which is thicker and prepared by hand from ground beef. Generally, most American hamburgers are round, but some fast-food chains, such as Wendy's, sell square-cut hamburgers. Hamburgers in fast food restaurants are usually grilled on a flat-top, but some firms, such as Burger King, use a gas flame grilling process. At conventional American restaurants, hamburgers may be ordered "rare", but normally are served medium-well or well-done for food safety reasons. Fast food restaurants do not usually offer this option.

The McDonald's fast-food chain sells the Big Mac, one of the world's top selling hamburgers, with an estimated 550 million sold annually in the United States. Other major fast-food chains, including Burger King (also known as Hungry Jack's in Australia), A&W, Culver's, Whataburger, Carl's Jr. /Hardee's chain, Wendy's (known for their square patties), Jack in the

Box, Cook Out, Harvey's, Shake Shack, In-N-Out Burger, Five Guys, Fatburger, Vera's, Burgerville, Back Yard Burgers, Lick's Homeburger, Roy Rogers, Smashburger, Taco Bell and Sonic also rely heavily on hamburger sales. Fuddruckers and Red Robin are hamburger chains that specialize in the mid-tier "restaurant-style" variety of hamburgers.

Some North American establishments offer a unique take on the hamburger beyond what is offered in fast food restaurants, using upscale ingredients such as sirloin or other steak along with a variety of different cheeses, toppings, and sauces. Some examples would be the Bobby's Burger Palace chain founded by well-known chef and Food Network star Bobby Flay.

Hamburgers are often served as a fast dinner, picnic or party food, and are usually cooked outdoors on barbecue grills.

(Hamburger, http://en.wikipedia.org/wiki/Hamburger)

Project 8
Some Bread Ingredients

Task 1 任务一：
认识各种主要原料名称，并能用英语表达出来。（采用小组讨论形式，最后用英语回答教师的问题）

Task 2 任务二：
说出每种原料的特征及作用，并掌握关键的英语表达。（采用小组讨论形式，最后用英语在讲台上表达）

Task 3 任务三：
用英语说出和写出原料名称及其在各种制作配方中的数量。（采用小组讨论形式，最后用英语在讲台上表达，知道最多的为本节课的优胜组）

Task 4 任务四：
大家讨论一下，在实训室里，哪种原料用在哪些产品中？（采用小组讨论形式，最后用英语在讲台上表达）

Which do you like most?

Some Bread Ingredients Project 8 项目八

Flour

The flour you choose for your bread also makes a difference in the quality of the final product. Bread flour makes a superior loaf. This flour is higher in protein content, and protein or gluten is what gives bread its unique texture.

When water is added to flour, two proteins, glutenin and gliadin, combine to form gluten. Gluten forms a network of proteins that stretch through the dough like a web, trapping air bubbles that form as the yeast ferments. This creates the characteristic air holes of perfect bread.

All purpose flour will also work just fine in most bread recipes. Don't use cake flour because there isn't enough protein in that type, and your bread will fall because the structure won't be able to withstand the pressure of the gasses the yeast creates.

Whole grain flours and other types of flour add color, texture, and flavor to breads. These flour types don't have enough gluten to make a successful loaf on their own, so all purpose or bread flour is almost always added to provide structure. Yeast:

Yeast

Make sure your yeast is fresh. Active dry yeast, sold in individual packets, is the easiest type to use, and keeps well in your pantry. There is always a "best if used by" date on the packages, and you should follow this rigorously. If you are going to take the time to make bread, fresh yeast is essential.

The temperature of the water, whether used to dissolve the yeast, or added to a yeast/flour mixture, is critical. Until you get some experience, use a thermometer. When the yeast is dissolved in the water or other liquid, the temperature must be 110 to 115 degrees. When the yeast is combined with flour and other dry ingredients, the liquid temperature can be higher, about 120 to 130 degrees.

Liquid

The type of liquid you use will change the bread characteristics. Water will make a loaf that has more wheat flavor and a crisper crust. Milk and cream-based bread are richer, with a finer texture. These bread brown more quickly because of the additional sugar and butterfat added to the dough. Orange juice is a nice addition to whole wheat bread because its sweetness helps counter the stronger flavor of the whole grain.

Salt

Salt is essential to every bread recipe. It helps control yeast development, and prevents the bread from over rising. This contributes to good texture. Salt also adds flavor to

the bread. It is possible to make salt-free bread, but other ingredients like vinegar or yogurt are added to help control the growth of yeast.

Fats

Fats like oils, butter and shortening add tenderness and flavor to bread. Bread made with these ingredients are also moister. Make sure you don't use whipped butter or margarine, or low fat products, since they contain water. The composition of the dough will be weakened, and your loaf will fail.

(Linda Larsen, *Bread Ingredients*, http://busycooks.about.com/od/howtobake/a/bread101.htm)

New Words and Expressions

1. make a difference 有影响；起（重要）作用
2. superior [suː'pɪərɪə(r)] *adj.* （在质量等方面）较好的
3. protein content ['prəʊtiːn 'kɒntent] 蛋白质含量
4. unique [juː'niːk] *adj.* 唯一的，仅有的；独特的
5. texture ['tekstʃə(r)] *n.* 质地；结构；本质
6. glutenin ['gluːtənɪn] *n.* 麦谷蛋白
7. gliadin ['ɡlaɪədɪn] *n.* 麦胶蛋白
8. combine [kəm'baɪn] *vt.* 使结合
9. form [fɔːm] *vt.* 形成，构成
10. network ['netwɜːk] *n.* 网；（电视与计算机）网络；网状物
11. stretch [stretʃ] *n.* 伸展，弹性；*vt.* 伸展；张开
12. web [web] *n.* 蜘蛛网，网状物
13. trap [træp] *vt.* 诱骗；困住
14. bubble ['bʌbl] *n.* 泡，水泡；*vt. & vi.* 冒泡，起泡
15. yeast ferment 酵母发酵
16. create [kriː'eɪt] *vt.* 创作；产生
17. characteristic [ˌkærəktə'rɪstɪk] *adj.* 特有的；独特的；*n.* 特性，特征，特色
18. hole [həʊl] *n.* 洞，孔；洞穴
19. perfect ['pɜːfɪkt] *adj.* 完美的；正确的；精通的
20. all purpose flour 通用粉
21. recipe ['resəpi] *n.* 食谱；处方；配方
22. fall [fɔːl] *vi.* 掉下，跌倒；倒塌，崩溃
23. structure ['strʌktʃə(r)] *n.* 结构；构造；建筑物
24. withstand [wɪð'stænd] *vt.* 经受，承受；耐得住，禁得起
25. pressure ['preʃə(r)] *n.* 压力；气压
26. whole grain flour 全麦面粉
27. successful [sək'sesfl] *adj.* 成功的
28. provide [prə'vaɪd] *vt. & vi.* 提供，供给，供应

Some Bread Ingredients Project 8 项目八

29. fresh ［freʃ］ *adj.* 新鲜的；淡水的；新的
30. active dry yeast ［'æktiv drai ji:st］活性干酵母
31. individual ［ˌindi'vidʒuəl］ *adj.* 个人的；独特的；个别的
32. packet ［'pækit］ *n.* 小包；信息包
33. pantry ［'pæntri］ *n.* 餐具室，食品储存室
34. rigorously ［'rigərəsli］ *adv.* 严厉地，残酷地；严密地
35. essential ［i'senʃl］ *adj.* 基本的；必要的
36. temperature ［'temprətʃə(r)］ *n.* 温度；体温
37. critical ［'kritikl］ *adj.* 批评的；决定性的，关键的
38. experience ［ik'spiəriəns］ *n.* 经验，体验；经历
39. thermometer ［θə'mɔmitə(r)］ *n.* 温度计；体温表
40. liquid ［'likwid］ *adj.* 液体的；清澈的；*n.* 液体，流体
41. degree ［di'gri:］ *n.* （数）度，度数；程度
42. crisper crust 脆皮
43. cream-based bread 奶油面包
44. brown ［braun］ *vt. & vi.* （使）呈（或变成）褐色（或棕色）（尤指日晒或经烘烤）；炸（或烤）成褐色
45. additional ［ə'diʃənl］ *adj.* 额外的，附加的；另外的
46. butterfat ［'bʌtəfæt］ *n.* 乳脂
47. orange juice 橙汁
48. sweetness ［'swi:tnəs］ *n.* 甜蜜；美妙；芳香
49. counter ［'kauntə］ *n.* 柜台；对立面；计数器；*vt.* 反击，还击；*vi.* 逆向移动，对着干；
50. prevent...from... ［pri'vent frɔm］阻止，防止
51. over rising ［'əuvə 'raiziŋ］过度膨胀
52. contribute to ［kən'tribjut tu:］有助于，贡献；为……做贡献
53. salt-free ［'sɔ:lt-fri:］无盐的
54. vinegar ［'vinigə(r)］ *n.* 醋
55. yogurt ［'jəugət］ *n.* 酸奶；酸乳酪
56. tenderness ［'tendənis］ *n.* 柔软；温和
57. moist ［'mɔist］ *adj.* 潮湿的，微湿的
58. whip ［wip］ *v.* 抽打；搅拌……直至变稠
59. low fat product 低脂肪产品
60. contain ［kən'tein］ *vt.* 包含，容纳
61. composition ［ˌkɔmpə'ziʃn］ *n.* 创作；构图，布置
62. weaken ［'wi:kən］ *vt. & vi.* （使）削弱；（使）变弱；衰减
63. fail ［feil］ *vt. & vi.* 失败，不及格

Grammar

Be going to 句型

表示"近期"或"打算"要做的事情时,往往用 be going to 句型。例如:

1. It is going to rain tomorrow.
   明天会下雨。
2. We are going to have an English class this afternoon.
   今天下午我们有英语课。
3. A light rain is going to fall on Zhengzhou this afternoon.
   今天下午郑州将下小雨。
4. There is going to be a sandstorm in Beijing.
   北京将有一场沙尘暴。
5. Are you going to see the film with us?
   你和我们一起去看电影吗?
6. What is Jane going to do?
   Jane 要去做什么?

Let's have a try!

Fill in the blanks with the proper form of the words in the brackets.

1. I _____ (visit) my grandparents next Sunday.
2. They _____ (leave) home at 8:00 tomorrow morning.
3. I _____ (see) her tomorrow.
4. She _____ (go) out in a few minutes.
5. I _____ (post) the letter this afternoon.
6. There _____ (be) a singing competition this evening.

Exercises

❶ "说一说"(用英语)
请说明每一种原料的名称及作用。

❷ "想一想"
实验的时候还会用到哪些原料呢?
把它们写下来。

Supplementary Knowledge

Special Classifications of Wheat Flours

Bread flour（面包粉）: Contains 13.5% ~ 14% gluten.

Cake flour（蛋糕粉）: Low in gluten.

Whole wheat flour（全麦粉）: Whole wheat flour is ground with the "whole" kernel—endosperm, germ and bran. However, whole wheat flours I've bought here in China seem less "whole". The color is not as dark, and I suspect that some of the bran seems to have been removed to give it a more pleasing flavor and texture. I would guess it's more similar to a T 90 or a T110 French flour than King Arthur whole wheat flour.

Dumpling flour（饺子粉）: A relatively high gluten flour (around 11%). If you can't find bread flour or high-gluten flour, dumpling flour is a good substitute as it is commonly sold in many supermarkets. (In fact, it is really about the only flour I've seen carried everywhere.)

Youtiao flour（油条粉）: This flour contains several chemical additives needed to create the porous, chewy, spongy texture of youtiao（油条）, deep-fried dough crullers. I haven't tried this flour, and don't really intend to, either.

Self-rising flour（自发粉）: A mid-gluten, all-purpose flour mixed with baking soda (sodium bicarbonate), and an acidic salt, sometimes monocalcium phosphate (which is found in baking powder as well).

Cheng fen（澄粉）: Wheat flour from which the gluten has been completely removed. It is essentially wheat starch, and is used in making sticky rice cakes and mochi for that distinctive chewy, glutinous quality.

"Wheat heart flour"（麦心粉）: Milled from the endosperm of the wheat kernel.

(*Baking Bread in China: A Flour Glossary*, http://www.hawberry.net/baking-bread-china-guide-ingredients-supplies/flour-guide/)

Module 2 到此结束！

Module 3:
The Process of Bread Production and Dough Processing Methods

模块三

面包生产主要工艺及生产方法

You will Learn:

- Dough Processing Methods
- Dough Mixing
- Baking Reaction

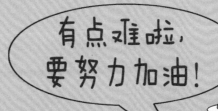

有点难啦,要努力加油!

Project 9
Dough Development During Mixing

Task 1 任务一：
了解面团搅拌的几个阶段。（采用小组讨论形式，最后用英语回答教师的问题）

Task 2 任务二：
掌握面团搅拌各个阶段的特征及注意事项，并掌握关键步骤的英语表达。（采用小组讨论形式，最后用英语在讲台上表达）

Task 3 任务三：
掌握面团搅拌到哪个阶段为最佳阶段。（采用小组讨论形式，最后用英语在讲台上表达）

Task 4 任务四：
大家讨论一下，我们在实训室里是怎样做的呢？（采用小组讨论形式，最后用英语在讲台上表达）

Mixing to the correct degree is of critical importance for the eventual behavior of the dough during subsequent processing, and for the ultimate quality of the final bread.

There are six stages during dough mixing.

The first stage: Dough pick-up. This is the first stage where the dry and wet materials have mixed only to the point of combining together. The mass is wet, rough and uneven.

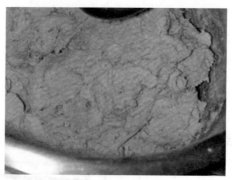

Stage 1: Dough pick-up

The second stage: Dough clean-up. In this stage the water is being taken up by the dry material, and forms a staff mass that is wet in spots. The back of the bowl is left clean.

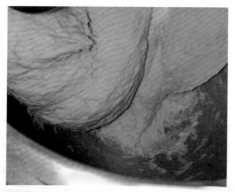

Stage 2: Dough clean-up

The third stage: Dough development. In this stage the dough commences to show life, it is becoming elastic and dryness is evident.

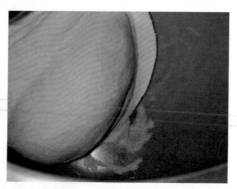

Stage 3: Dough development

The fourth stage: Dough final. At this stage the dough has reached its peak of development. It has a silky and dry appearance. No evidence of wetness or roughness. The dough is smooth and mellow and gives a long stretch.

Stage 4: Dough final

The fifth stage: Dough letdown. At this stage the dough is forming thin strings that are wet and sticky. Unless the flour is tolerant, this is a dangerous zone to mixing operation.

Stage 5: Dough letdown

The sixth stage: Here the dough has broken down and disintegrated. The mechanical punishment is severe and the dough has liquefied. It is slack and translucent.

Stage 6: Dough broken down

(Excerpt from E. J. Pyler, *Baking Science and Technology*)

New Words and Expressions

1. the correct degree 恰当的程度
2. subsequent ['sʌbsikwənt] adj. 接下来的，随后的
3. processing [prəʊ'sesiŋ] n. 处理，加工
4. ultimate ['ʌltimət] adj. 最后的
5. rough and uneven 粗糙不平
6. clean up 卷起
7. take up 吸收
8. staff mass [stɑːf mæs] 松散的一团
9. in spots 有些地方
10. dough development 面筋开始扩展阶段
11. commence [kə'mens] vt. 开始，着手
12. elastic [i'læstik] adj. 有弹力的，可伸缩的
13. dryness ['drainis] n. 干燥
14. evident ['evidənt] adj. 明显的，清楚的
15. dough final 面筋扩展完成阶段
16. peak of development 面筋扩展最佳状态
17. silky ['silki] adj. 光滑的
18. appearance [ə'piərəns] n. 外表
19. smooth and mellow 圆润光滑
20. dough letdown ['let daun] 面筋搅拌过度
21. string [striŋ] n. 线丝，植物纤维
22. sticky ['stiki] adj. 黏性的
23. tolerant ['tɔlərənt] adj. 可容忍的
24. dangerous zone ['deindʒərəs zəun] 危险区域
25. dough broken down 面筋打断阶段
26. disintegrated [dis'intigreitid] vi. 破裂的，分裂的
27. mechanical punishment [mə'kænikl 'pʌniʃmənt] 机械失误后果
28. liquefied ['likwifaid] adj. 液化的，溶解的
29. slack [slæk] adj. 松弛的
30. translucent [træns'luːsnt] adj. 半透明的，透亮的，有光泽的

Grammar

连词的用法

连接句子或句子中相同的成分时用连词，如：and（和，以及），but（但是）以及 or（或者）。例如：

1. I meet different people every day, old and young, men and women.

2. I like meat but he likes vegetables.

3. Would you like to change for another one or just get the money back?

Fill in the blanks with prepositions.

1. Do you like ice cream _____ coke?
2. I bought some apples _____ oranges.
3. I hope to become a tour guide, _____ I don't want to work long hours.
4. Do you want to be a cashier _____ a factory worker?
5. The work is tiring _____ interesting.

❶ "想一想"
打面团共有几个步骤，请把每个步骤的中英文写下来。

❷ "议一议"
你们搅拌面团是搅拌到哪个阶段？
怎样判断面筋已经打好？

❸ "试一试"
写出面团在每个步骤显现的特征：

Supplementary Knowledge

The History of Bread

Bread, in one form or another, has been one of the principal forms of food for man from earliest times.

The trade of the baker, then, is one of the oldest crafts in the world. Loaves and rolls have been found in ancient Egyptian tombs. In the British Museum's Egyptian galleries you can see actual loaves which were made and baked over 5,000 years ago. Also on display are grains of  wheat which ripened in those ancient summers under the Pharaohs. Wheat has been found in pits where human settlements flourished 8,000 years ago. Bread, both leavened and unleavened, is mentioned in *the Bible* many times. The ancient Greeks and Romans knew bread for a staple food even in those days people argued whether white or brown bread was the best.

Further back, in the Stone Age, people made solid cakes from stone-crushed barley and wheat. A millstone used for grinding corn has been found, that is thought to be 7,500 years old. The ability to sow and reap cereals may be one of the chief causes which led man to dwell in communities, rather than to live a wandering life hunting and herding cattle.

According to botanists, wheat, oats, barley and other grains belong to the order of grasses, nobody has yet found the wild form of grass from which wheat, as we know it, has developed. Like most of the wild grasses, cereal blossoms bear both male and female elements. The young plants are provided with a store of food to ensure their support during the period of germination; and it is in this store of reserve substance that man finds an abundant supply of food.

(*History of bread*, http://www.botham.co.uk/bread/history1.htm)

Project 10
Chiffon Cakes

Task 1 任务一：
了解戚风蛋糕起发的主要途径。（采用小组讨论形式，最后用英语回答教师的问题）

Task 2 任务二：
戚风蛋糕所用的油一般是什么油？（采用小组讨论形式，最后用英语在讲台上表达）

Task 3 任务三：
戚风蛋糕的蛋白部分应搅拌到哪个阶段，蛋白搅拌的湿性发泡有什么特征？（采用小组讨论形式，最后把一些英语关键词写在黑板上，写出最多的为本节课的优胜队）

Task 4 任务四：
大家讨论一下，我们在实训室里是怎样做戚风蛋糕的呢？（采用小组讨论形式，最后用英语在讲台上表达）

Chiffon cakes resemble Angel Food cakes in that their leavening depends principally on the whipping of egg whites. Both types of cake utilize the same basic ingredients, with the exception that chiffon cakes additionally contain egg yolks and fat, the latter normally in the form of a vegetable oil, such as a good grade of cottonseed oil. A typical chiffon cakes formula is given in the following table.

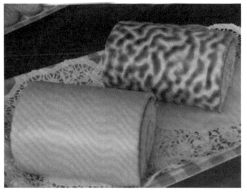

Chiffon Cake's Formula

Ingredients	Lbs.
Egg yolk	10.75
Oil	10.75
Egg whites	32
Salt	0.375
Granulated sugar	16
Cake flour	14
Powdered sugar	16
Cream of tartar	0.125
Total weight	100.00

  The order in which the ingredients are combined and the mixing procedure itself are of considerable importance. Two separate mixing bowls are required, one for whipping the egg white and salt into a soft foam, followed by the addition of the cream of tartar and granulated sugar with continued whipping, and the other for whipping the egg yolks with slow addition of the oil until a uniform, aerated mixture is obtained. The yolk and oil mixture is then carefully poured into the egg white foam. The flour and sugar are sifted together and then blended into the egg mixture.

(Excerpt from E. J. Pyler, *Baking Science and Technology*)

New Words and Expressions

1. utilize ['juːtlaɪz] *vt.* 利用，使用
2. exception [ikˈsepʃn] *n.* 例外
3. additionally [əˈdɪʃənəli] *adv.* 额外地
4. the latter 后者
5. normally [ˈnɔːməli] *adv.* 通常地
6. cottonseed oil [ˈkɔtənsiːd ɔil] 棉籽油
7. typical chiffon cakes formula 有代表性的戚风蛋糕配方
8. considerable [kənˈsidərəbl] *adj.* 相当的
9. soft foam [sɔft fəum] 湿性发泡
10. cream of tartar （有机化学）酒石，酒石酸氢钾
11. granulated sugar [ˈgrænjuleitid ˈʃugə(r)] 砂糖
12. uniform [ˈjuːnifɔːm] *adj.* 均衡的
13. aerated [ˈeiəreitid] *adj.* 充气的
14. pour into 慢慢倒入
15. egg white foam 蛋白泡沫
16. blend into [blend ˈintuː] 融入，与……融合

Grammar

Such as 的基本用法

1. 表示举例

意为"例如，诸如此类的，像……那样的"，相当于 like 或 for example。如：There are few poets such as Keats and Shelly. 像济慈和雪莱这样的诗人现在很少了。

2. 表示"像……这样的"

其中的 as 用作关系代词，引导定语从句，as 在定语从句中用作主语或宾语。此外，不要按汉语意思把该结构中的 as 换成 like。如：

He is not such a fool as he looks. 他并不像他看起来那么傻。

3. 表示"凡是……的人（或事物）"、"所有……的人（事物）"

其意相当于 everything that, all those, those that (who) 等。其中的 such 为先行词，as 为关系代词。如：

Take such (things) as you need. 你需要什么就拿什么。

You may choose such as you prefer. 你可挑选自己想要的东西。

4. 用作关系代词

有时 such as 整个用作关系代词，用以引导定语从句。如：

We had hoped to give you a chance such as nobody else ever had.

我们本来希望给你一个别人从未有过的机会。

Let's have a try!

Complete the following sentences.

1. I've never heard _____. 我从未听过他讲那样的故事。
2. He is _____ we all respect. 他是一位我们大家都尊敬的好老师。
3. I enjoy _____ this one. 我喜欢像这首歌一样的歌。
4. I know four languages, _____. 我懂四种语言，如日语、英语等。
5. They planted many flowers, _____ sunflowers, etc. 他们种了许多种花，如玫瑰花、向日葵等。

Exercises

❶ "做一做"
请用中文列出该戚风蛋糕所需的原料及数量。

❷ "试一试"
请用中文简单描述戚风蛋糕的制作程序或步骤：

Supplementary Knowledge

History of Cakes

Cakes are made from various combinations of refined flour, some form of shortening, sweetening, eggs, milk, leavening agent, and flavoring. There are literally thousands of cakes recipes (some are bread-like and some rich and elaborate) and many are centuries old. Cake making is no longer a complicated procedure.

Baking utensils and directions have been so perfected and simplified that even the amateur cook may easily become an expert baker. There are five basic types of cake, depending on the substance used for leavening.

The most primitive peoples in the world began making cakes shortly after they discovered flour. In medieval England, the cakes that were described in writings were not cakes in the conventional sense. They were described as flour-based sweet foods as opposed to the description of bread which were just flour-based foods without sweetening.

Bread and cake were somewhat interchangeable words with the term "cake" being used for smaller bread. The earliest examples were found among the remains of Neolithic villages where archaeologists discovered simple cake made from crushed grains, moistened, compacted and probably cooked on a hot stone. Today's version of this early cake would be oatcake, though now we think of it more as a biscuit or cookie.

Cakes were called "plakous" by the Greeks, from the word "flat". These cakes were usually combinations of nuts and honey. They also had a cake called "satura", which was a flat heavy cake.

During the Roman period, the name for cake (derived from the Greek term) became "placenta". They were also called "libum" by the Romans, and were primarily used as an offering to their gods. Placenta was more like a cheese cake, baked on a pastry base, or sometimes inside a pastry case.

The terms "bread" and "cake" became interchangeable as years went by. The words themselves are of Anglo Saxon origin, and it's probable that the term cake was used for the smaller bread. Cakes were usually baked for special occasions because they were made with the finest and most expensive ingredients available to the cook. The wealthier you were, the more likely you might consume cake on a more frequent basis.

By the middle of the 18th century, yeast had fallen into disuse as a raising agent for cakes in favor of beaten eggs. Once as much air as possible had been beaten in, the mixture would be poured into molds, often very elaborate creations; but sometimes as simple as two tin hoops, set on parchment paper on a cookie sheet. It is from these cake hoops that our modern cake pans developed.

Cakes were considered a symbol of well-being by early American cooks on the east coast, with each region of the country having their own favorites.

By the early 19th century, due to the Industrial Revolution, baking ingredients became more affordable and readily available because of mass production and the railroads. Modern leavening agents, such as baking soda and baking powder were invented.

(*History of Cakes*, http://whatscookingamerica.net/History/CakeHistory.htm)

Project 11
Dough Processing Methods

Task 1 任务一：
用英语说出面包生产的 3 种方法。（采用小组派代表到讲台用英语表达的形式，说得最准确、最迅速的为本节课的优胜组）

Task 2 任务二：
用英语说出使用一次发酵法时，面团搅拌后的温度范围。（各小组派代表到讲台用英语表达，说得最准确、最迅速的可得奖励）

Task 3 任务三：
根据课文内容，讨论在中种发酵法中，面种的面粉用量与面种的发酵时间之间是怎样的关系？用英语说出其关键词。（采用小组讨论形式，最后把一些英语关键词在黑板上写出来，写出最多的为本节课的优胜组）

Task 4 任务四：
大家讨论一下，我们在实训室里都用过哪种面包生产方法呢？（采用小组讨论形式，最后用英语在讲台上表达）

Which do you like most?

1. Straight Dough Method

The straight dough method is a single-step process in which all the ingredients are mixed together into a single batch. Mixing in this case is continued until the dough assumes a smooth appearance and acquires an optimum elastic character. The temperature of the dough out of mixed should be within the range of 78°F ~ 82°F (25.5℃ ~ 28℃). Higher temperatures are ordinarily intended to reduce the fermentation time, although for this purpose the use of somewhat greater than normal amounts of yeast is preferable. Fermentation is conducted for two to four hours—the actual time in practice being normally close to three hours—after which the dough is divided and made up in a manner essentially identical to that employed with sponge-dough.

Figure 1 Before fermentation

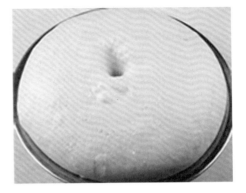
Figure 2 After fermentation

2. Sponge and Dough Method

In the sponge and dough method, the major fermentative action takes place in a pre-ferment, which is referred to as a sponge, in which normally more than one half of the total dough flour is subjected to the physical, chemical and biological action of an active yeast fermentation.

After the sponge has been mixed, it is set to ferment until it has reached the proper degree of maturity or ripeness which is indicated by perceptible drop in the sponge volume, usually called the break. This may require a fermentation time of 3.5 hours, in the case of a sponge incorporating 75 percent of total flour, and a fermentation time of about five hours in a sponge made up of 50 percent of the total flour.

Figure 3 The sponge after fermentation

The second step of the sponge and dough method is the so-called dough stage in which the fermented sponge is returned to the mixer and the rest of the dough ingredients are added. These

comprise the balance of the materials required by the formula and include the remaining flour and water, and normally the milk solids, salt, shortening and any other additional ingredients called for by the formula.

The dough is mixed to its optimum condition and then returned to the fermentation room where it is given a "floor time", or second fermentation, for a period ranging anywhere from 15 minutes to about one hour. Because the sponge portion of the dough is subjected to double mixing, namely, once in the sponge stage and then again in the dough stage.

3. No-Time Dough Method

The use of no-time dough in a bakery is generally an emergency measure, when the production does not meet the sales demand.

The dough is usually made with a greater percentage of yeast, based on a straight dough formulation. Often the dough temperatures are increased, as are the temperatures and humidity during the proofing stage.

The dough, after the normal mixing period, is given up to half an hour floor time, or rest period, during which time visible signs of fermentation become evident, due to the high yeast level used.

At this point the dough is scaled, rounded, and molded, following the more usual procedures.

(Excerpt from E. J. Pyler, *Baking Science and Technology*,
and Sylvia M. J. Enkins, *Bakery Technology*)

New Words and Expressions

1. straight dough method 直接发酵法，一次发酵法
   straight [streit] *adv.* 直接地
2. a single-step process 一次发酵工艺
   process [ˈprəuses] *n.* 过程；工序；做事方法；工艺流程
3. a single batch 单个面团
4. assume [əˈsjuːm] *vt.* 呈现
5. smooth appearance 光滑表面
6. acquire [əˈkwaiə] *vt.* 获得
7. an optimum elastic character 最佳弹性
   optimum [ˈɔptiməm] *adj.* 最适宜的 *n.* 最佳效果；最适宜条件；（生物学）最适度
   character [ˈkærəktə] *n.* 性格，品质；特征
8. ordinarily [ˈɔːdinərili] *adv.* 平常地；通常地；一般地；按说
9. reduce [riˈdjuːs] *vi.* 减少
10. preferable [ˈprefərəbl] *adj.* 更好的，更可取的
11. normal amounts 正常量

12. conduct [kən'dʌkt] vt. & vi. 引导，带领；控制；传导；组织，安排；实施
13. practice ['præktis] n. 实践
14. close to 近乎；临近
15. essentially [i'senʃəli] adv. 本质上，根本上
16. identical to [ai'dentikəl tuː] 与……相同
17. employ [im'plɔi] vt. 使用，利用
18. sponge-dough 发酵面团，中种面团，面种（专指面包制作中第一次发酵的面团）
19. sponge and dough method 二次发酵法，中种发酵法
20. action ['ækʃ(ə)n] n. 反应
21. take place 发生
22. pre-ferment 第一次发酵
23. refer to 指的是，提及，意指
24. subject to ['sʌbdʒikt tuː] 受限于；以……为准；易受……影响
25. active yeast formation 活性酵母发酵
26. proper degree 合适的程度
27. maturity [mə'tʃuərəti] n. 成熟
28. ripeness ['raipnəs] n. 成熟，老练；成熟度
29. indicate ['indikeit] vt. 表明，标示
30. perceptible [pə'septəbl] adj. 可感觉（感受）到的，可理解的，可认识的
31. sponge volume 面团的体积
32. incorporate [in'kɔːpəreit] vt. 包含；使混合
33. so-called adj. 所谓的，号称的
34. dough stage 主面团阶段
35. the fermented sponge 发酵后的中种面团
36. return to 再放回……
37. comprise [kəm'praiz] vt. 包含，包括；由……组成；由……构成
38. balance ['bæləns] n. 余额
39. remaining [ri'meiniŋ] adj. 剩下的；剩余的
40. milk solids 奶粉
    solid ['sɔlid] adj. 固体的；实心的 n. 固体
41. optimum condition 最佳状态；最适宜条件
    condition [kən'diʃn] n. 情况，状态；环境，条件
42. floor time 延续发酵，案台发酵
    floor [flɔː(r)] n. 地面，地板；楼层
43. second fermentation 二次发酵

Learning Tips

英语和美语在读音上的差异：

3. 在以 -ile 结尾的一类单词中，英国人将尾音节中的字母 i 读作长音 [ai]；而美国人则弱读作 [ə]，例如：

	英语读音	美语读音
docile	['dousail]	['dɔsəl]
fertile	['fɜːtail]	['fɜːtl]

44. double mixing 两次搅拌

    double ['dʌbl] adj. 双的；两倍的

45. no-time dough method 快速发酵法

46. emergency measure 紧急措施

    emergency [i'mɜːdʒənsi] adj. 紧急的，应急的

    measure ['meʒə(r)] n. 测量，测度；措施

47. meet the sales demand 满足销售需求

    demand [di'mɑːnd] n. 需求

48. percentage [pə'sentɪdʒ] n. 百分比

49. base on 在……的基础上

50. the proofing stage 醒发阶段

51. up to 达到

52. rest period 醒发时间

53. visible sign 可见的迹象

    visible ['vɪzəbl] adj. 看得见的；明显的

    sign [saɪn] n. 记号；预兆

54. due to 由于

Grammar

如何表示"频率"

当我们描述一个动作发生的频率的时候，可以用这些词：always, usually, often, sometimes, seldom, never。例如：

1. We always have classes in the morning.

   我们总在上午上课。

2. In the afternoon, we usually go outside of the school.

   下午我们通常到校外去。

3. He often plays basketball after school.

   他经常在放学后打篮球。

4. Sometimes we work as tour guides.

   有时我们做导游。

5. He seldom has supper after 8:00 in the evening.

   他很少在晚上8点之后吃晚饭。

6. She is never late for school.

   她上学从不迟到。

Dough Processing Methods Project 11 项目十一

Complete the sentences according to the Chinese in the brackets.

1. She _____ （从不） gets up before 7 o'clock.
2. He _____ （经常） has classes in the morning.
3. They _____ （很少） visit factories in the afternoon.
4. _____ （有时） I have breakfast on my way to school.

Exercises

我们来做做下面的练习。

❶ "找一找"
文章里介绍了哪几种面包生产方法？列举出相对应的中英文名称。

❷ "说一说"
详述每一种面包生产方法的步骤与特点。

❸ "想一想"
你们搅拌面团通常搅拌到哪个阶段？怎么判断面筋已经打好了？

Supplementary Knowledge

Top 10 Types of Bread That Are Healthy

The bread section is the most popular section at the grocery store and certainly the most confusing one. The most important aspect to look for in the bread is the ingredients. Make sure to choose bread which is 100% whole grain. Whole grains are a great source of fiber, which helps in

the prevention of heart disease and colon cancer. Fiber also helps to keep us full for a long time so that we eat less throughout the day and maintain our weight. Below are some of the healthiest types of bread.

1. **Flaxseed Bread**: Bread that is prepared with flaxseed is very crispy to eat and has lots of good cholesterol that keeps the heart healthy. Flaxseed is also the only vegetarian source of omega-3 fatty acids, hence vegetarians must definitely go for this one.

2. **Pita Bread**: The pita bread is usually not counted among nutritional bread. But it is a part of the Mediterranean diet that is known to be heart healthy. It is said that people who follow this diet have less chances of developing heart disorders.

3. **Gluten-free Bread**: Gluten is a protein found in wheat, barley and rye. Gluten-free bread are generally made with grains other than wheat to make them gluten free. Whole grain gluten free bread may be made with brown rice flour.

4. **Pumpernickel**: Pumpernickel is basically a kind of rye bread. But along with rye, it also contains soy and other whole grains. That is why pumpernickel bread is one of the healthiest varieties of bread.

5. **Walnut Bread**: Walnuts are one of the nutritious nuts in the world. Walnuts are rich in omega-3 fatty acids. Thus, the bread that is prepared with walnut carries lots of good cholesterol for the heart.

6. **Oat Bread**: Oat bread is one of the healthiest bread available to us. Oats are very heart-healthy because they absorb bad cholesterol. Bread made from oats are highly rich in fiber and healthy for the heart, as well.

7. **Rye Bread**: Rye bread is one of the healthiest varieties of bread available in the market. But make sure it actually contains rye and also has the taste of rye in it.

8. **Multi-grain Bread**: Multi-grain bread is a combination of various grains like wheat, rye, millet, etc. That is why it is rich in nutritious fibers and thus heart-healthy.

9. **Brown Bread**: Brown bread is usually a combination of wheat and soy that gives it the rich brown color. Brown bread is heart-healthy mainly because it is low-fat bread and also gluten-free.

10. **Whole Wheat Bread**: Whole wheat bread is made from unrefined and unpolished wheat. That is why it is healthy, fibrous and also good for the heart. Whole wheat, being rich in fiber, is a good choice of bread.

Hopefully, this article proves to be helpful for the readers in choosing the right type of bread that keeps the heart healthy.

（选自"大耳朵英语"网站，http：//www.bigear.cn/news-65-103684.html）

Project 12
Baking Reaction

Task 1 任务一：
用英语说出你使用过或见过的烤炉种类名称。（采用小组派代表到讲台用英语表达的形式，说得最准、最快的为本节课的优胜组）

Task 2 任务二：
根据课文内容，用英语说出 1 个面团中有多少个微小的气室？（采用小组讨论形式，最后把一些英语关键词在黑板上写出来，写得最多、最准的为本节课的优胜组）

Task 3 任务三：
根据课文内容，讨论面团入炉后的三种热物理效应。（各小组派代表到讲台用中文表达，说得最准确、最迅速的可得奖励）

Task 4 任务四：
大家讨论一下，烘焙急胀是怎么一回事呢？（采用小组讨论形式，最后用英语在讲台上表达）

Which do you like most?

When dough is placed in the oven, the first observable effect produced by the heat in the almost instantaneous formation of a thin, and at first readily expandable, surface film. The time span over which this initial film remains expandable will depend on the prevailing temperature and moisture conditions within the oven. For the first few minutes of oven time, the dough piece continues its regular progressive increase in volume which represents the so-called "oven rise".

The next important reaction is the sudden expansion of the dough volume by about one-third of original size, called "oven spring". The oven spring, while attributable to the effect of heat penetration, is actually the culmination of a whole series of reactions. A purely physical effect of heat upon a gas is to increase its pressure. If such a gas is

confined in an elastic or expandable container, the visible effect is an expansion of the container. A dough piece contains millions of minute gas cells in which the gas, under the influence of heat, begins to increase in pressure and causes the expansion of the confining cell walls.

Another purely physical effect of the heat is to reduce the solubility of gases. A considerable proportion of the carbon dioxide generated by the yeast is present in the dough in solution in the dough's liquids. Hence, as the temperature of the dough rises to about 120 °F (49℃), the carbon dioxide held in solution is liberated. This freed gas, rather than creating additional gas cells, migrates into the existing cells and adds to their interior pressure.

A third physical effect of heat is to change liquids with a low boiling point into vapors by the familiar process of distillation. Alcohol constitutes quantitatively the major low-boiling liquid in dough, so that it is transformed into vapor early in the baking process. This evaporation of alcohol at about 175 °F (79℃) in turn increases the gas pressure leading to an additional expansion of the gas cells.

(Excerpt from E. J. Pyler, *Baking Science and Technology*)

New Words and Expressions

1. baking reaction 烘焙反应
2. observable [əb'zɜːvəbl] *adj.* 观察得到的，可见的
3. effect [i'fekt] *n.* 反应，效果
4. instantaneous [ˌinstən'teiniəs] *adj.* 瞬间的，即时的
5. formation [fɔː'meiʃn] *n.* 形成
6. readily ['redili] *adv.* 容易地，无困难地；乐意地
7. surface film 表面薄膜
   surface ['sɜːfis] *n.* 表面
   film [film] *n.* 薄膜
8. expandable [ik'spændəbl] *adj.* 可膨胀的，可张开的
9. prevailing temperature 主温 prevailing [pri'veiliŋ] *adj.* 占优势的，盛行的
10. moisture condition 湿度条件
    moisture ['mɔistʃə(r)] *n.* 湿度
11. progressive increase 慢速膨胀
    progressive [prə'gresiv] *adj.* 渐进的
    increase [in'kriːs] *n. /vi. & vt.* 增长
12. in volume ['vɔljuːm] 体积上
13. oven rise 炉内膨胀
14. sudden expansion 突然膨胀

Learning Tips

我们在练习听力时，要注意一句话中的关键词，比如动词、形容词、名词等，它们一般含有该句的主要信息。听出这些词后，理解整句就会变得简单得多。例如 jammed, show, fix, open, push, remove 等词就很关键。

15. oven spring 烘焙急胀
16. attributable to [ə'tribjutəbl] 归功于……
17. heat penetration 热穿透
    penetration [peni'treiʃn] n. 渗透，突破
18. culmination [kʌlmi'neiʃn] n. 顶点，高潮
19. a whole series of reactions 一系列反应
20. purely physical effect 纯物理反应
    purely ['pjʊəli] adv. 纯粹地，完全地
    physical ['fizikl] adj. 物理的
21. container [kən'teinə(r)] n. 容器
22. gas [gæs] n. 气体
23. millions of 成千上万的，无数的
24. confine [kən'fain] vt. 限制
25. minute gas cell 微小的气室
    minute ['minit] adj. 微小的
    cell [sel] n. 细胞，小的空间
26. confining cell wall 受限的、狭小的、拘束的室壁
27. solubility [sɒljʊ'biləti] n. 溶解度
28. proportion [prə'pɔːʃn] n. 比例，部分；面积
29. carbon dioxide ['kɑːbən dai'ɒksaid] n. 二氧化碳
30. present ['preznt] adj. 现在的，目前的；出席的
31. in solution [sə'luːʃən] 溶解着；在不断变化中
32. liberated ['libəreitid] adj. 被解放的；思想解放的，无拘束的
33. freed gas 释放的气体
    freed [friːd] adj. 释放的
34. migrate into 迁移，移往
35. existing ['igzistiŋ] adj. 目前的；现存的
36. interior pressure 内压
    interior [in'tiəriə] adj. 内部的；室内的
37. boiling point ['bɔiliŋ pɔint] 沸点
38. vapor ['veipə] n. 水汽，水蒸气
39. distillation [disti'leiʃn] n. 蒸馏（过程）；蒸馏物
40. alcohol ['ælkəhɒl] n. 乙醇，酒精
41. constitute ['kɒnstitjuːt] vt. 构成，组成
42. quantitatively ['kwɒntətetivli] adv. 数量上
43. low-boiling liquid 低沸点液体
    low-boiling ['ləʊ'bɔiliŋ] adj. 低沸点的
44. transform into 把……转变成……

45. evaporation [i'væpə'reiʃn] n. 蒸发，汽化；消失，发散
46. in turn 相应地；转而
47. lead to [liːd] 通向；导致

Grammar

祈使句的用法

当你想请求或命令对方做某事时，可以使用祈使句结构。因为是直接对对方说，一般不用主语，直接用动词。如我们说 Sit down, please. 而不说 You sit down, please. 例如：

1. Look at the board.
2. Don't worry.
3. Pass me the book, please.

Complete the sentences with *come*, *give*, *pass*, *press*, *start* or *stop*.

1. Hi, Lisa. _____ on in, please.
2. Ken, _____ me the remote control, please.
3. My machine is out of order. _____ me a hand, please.
4. _____ the machine when you are ready to work.
5. _____ the UP button if you want to move it upward.
6. _____ the machine when something is wrong.

Exercises

"议一议"

请列举烘焙反应有哪几种，每一种反应是由什么引起的，各自的表现特征是什么？

Supplementary Knowledge

Baking Oven

A baking oven is a heating installation for baking bread products. It is the most important piece of equipment used in commercial baking. A baking oven consists of a source of heat, a baking chamber usually equipped with a steam humidifier, a conveyor with facilities for loading the dough and unloading the bread, an automatic regulating system for the baking process, and a device to recover the heat of lost gases. The average temperature in the baking chamber is 200℃ ~ 300℃, and the relative humidity is 15 ~ 70 percent. With respect to the nature of the working process and the equipment, baking ovens are similar to confectionery ovens.

Baking ovens are classified as batch or continuous ovens, depending on the method of operation, and as conventional and pass-through types (single-level and multilevel), depending on the design of the baking chamber. In conventional ovens the pieces of dough are loaded and the baked goods removed from the same side, in pass-through ovens the operations are performed from opposite sides. Ovens are also classified by the type of hearth. They may have cradle hearths suspended from a chain conveyor, plates mounted on a chain conveyor that form a continuous horizontal hearth, a lattice hearth in the form of a belt conveyor, a disk rotating about a vertical axis, a ring, or a draw plate or fixed sole hearth. Baking ovens are also classified according to the means used to heat the baking chamber. They may have tubular sections to which high-pressure saturated steam or superheated water is supplied, flat or tubular channels through which the combustion products of a fuel and recirculated gases are passed, or a baking chamber heated directly by electric heaters, gas burners, or infrared radiation lamps. Some use a combination of heating methods.

Baking ovens may be mechanized to various degrees. In automated ovens with conveyor hearths, the conveyor travel, heating conditions, and steam supply are automatically regulated, and the loading of dough pieces and unloading of the bread are mechanized or automated, provision is made for an automatic device to ensure safe fuel combustion. In mechanized ovens with conveyor hearths, the conveyor travel is regulated, the unloading of the finished goods is

mechanized, and provision is made for an automatic device to ensure safe fuel combustion. Other mechanized ovens have disk or extensible hearths, they are being supplanted by more modern designs. Nonmechanized ovens are used only in small enterprises.

The productivity of baking ovens depends on the design and may be as high as 100 ~ 120 tons per day.

(Excerpt from *The Great Soviet Encyclopedia*, http://encyclopedia2.thefreedictionary.com/Baking + Oven)

附录一
生词和词组汇总

A

acquire [əˈkwaiə] vt. 获得
action [ˈækʃ(ə)n] n. 反应
active dry yeast [ˈæktiv drai ˈjiːst] 活性干酵母
active yeast formation 活性酵母发酵
add [æd] vt. 增加，补充
addition [əˈdiʃn] n. 加，增加
additional [əˈdiʃənl] adj. 额外的，附加的；另外的
additionally [əˈdiʃənəli] adv. 额外地
aerated [ˈeiəreitid] adj. 充气的
alcohol [ˈælkəhɔl] n. 乙醇，酒精
all purpose flour 通用粉
an experienced baker 一位经验丰富的面包师
an optimum elastic character 最佳弹性
and so on 等等；诸如此类的；依此类推（用在诸多列举项目之后，相当于 etc.）
angel food cake [ˈeindʒəl fuːd keik] 天使蛋糕
appear [əˈpiə(r)] vi. 出现
appearance [əˈpiərəns] n. 外表
apply [əpˈlai] vt. 应用，涂；敷
approximately [əˈprɔksimətli] adv. 大约
a single batch 单个面团
a single-step process 一次发酵工艺
assume [əˈsjuːm] vt. 呈现
attractive [əˈtræktiv] adj. 有吸引力的
attributable to [əˈtribjutəbl] 归功于……
a whole series of reactions 一系列反应

B

bag out 裱花

baker ['beikə] n. 面包师；烤炉

baker's % 烘焙百分比（专门用于烘焙生产中）

bakery ['beikəri] n. 面包房，面包店

baking powder ['beikiŋ'paudə] n. 发酵粉，发粉

baking reaction 烘焙反应

balance ['bæləns] n. 余额

base on 在……的基础上

batter ['bætə] n. 面糊（用鸡蛋、牛奶、面粉等调成的糊状物）

be ready 准备好

belong to 属于；归于

Black Forest cake [blæk 'fɔrist keik] 黑森林蛋糕

blend into [blend 'intu:] 融入，与……融合

boiling point ['bɔiliŋ pɔint] 沸点

bowl [bəul] n. 搅拌缸

bread [bred] n. 面包

bread flour [bred 'flauə] 高筋面粉，面包专用面粉

bread-making process [bred 'meikiŋ 'prəuses] 面包制作过程

brown [braun] vt. & vi. （使）呈（或变成）褐色（或棕色）（尤指日晒或经烘烤）；炸（或烤）成褐色

brown sugar 红糖

bubble ['bʌbl] n. 泡，水泡；vt. & vi. 冒泡，起泡

butter ['bʌtə] n. 牛油（天然牛油）

butter cake 奶油蛋糕

butterfat ['bʌtəfæt] n. 乳脂

C

cake [keik] n. 蛋糕；糕饼

cake decoration [keik ˌdekə'reiʃən] 蛋糕装饰

carbon dioxide ['kɑ:bən dai'ɔksaid] n. 二氧化碳

cell [sel] n. 细胞，小的空间

change ['tʃeindʒ] n. 变化；找回的零钱（为不可数名词，不能加 s）

character ['kærəktə] n. 性格，品质；特征

characteristic [ˌkærəktə'ristik] adj. 特有的；独特的；n. 特性，特征，特色

charge [tʃɑ:dʒ] vi. 索价；收费

cheese cake [tʃi:z keik] 芝士蛋糕，奶酪蛋糕

chiffon cake ［'ʃifɔn keik］威风蛋糕
chocolate ［'tʃɔkəlit］n. 巧克力；adj. 用巧克力制的
chocolate cake 巧克力蛋糕
clean up 卷起
close to 近乎；临近
combine ［kəm'bain］vt. 使结合
commence ［kə'mens］vt. 开始，着手
composition ［kɒmpə'ziʃn］n. 创作；构图，布置
comprise ［kəm'praiz］vt. 包含，包括；由……组成；由……构成
condition ［kən'diʃn］n. 情况，状态；环境，条件
conduct ［kən'dʌkt］vt. & vi. 引导；带领；控制；传导；组织，安排；实施
confine ［kən'fain］vt. 限制
confining cell wall 受限的、狭小的、拘束的室壁
considerable ［kən'sidərəbl］adj. 相当的
constitute ［'kɔnstitjuːt］vt. 构成，组成
contain ［kən'tein］vt. 包含，容纳
container ［kən'teinə(r)］n. 容器
contribute to ［kən'tribjut tuː］有助于，贡献；为……做贡献
cookie ［'kuki］n. 曲奇饼
cool ［kuːl］vt. 冷却
corn starch ［kɔːn staːtʃ］n. 玉米淀粉
cost ［kɔst］vi. 价钱为；花费 n. 价钱；代价；花费
cottonseed oil ［'kɔtənsiːd ɔil］棉籽油
counter ［'kauntə］n. 柜台；对立面；计数器；vt. 反击，还击；vi. 逆向移动，对着干
cream ［kriːm］vt. 把……搅成糊状（或奶油状）混合物
cream filling 打发奶油，果酱，奶油馅
cream of tartar （有机化学）酒石，酒石酸氢钾
cream-based bread 奶油面包
create ［kri'eit］vt. 创作，产生
crisper crust 脆皮
critical ［'kritikl］adj. 批评的；决定性的，关键的
croissant ［'krwɑ'sɑː］n. 牛角酥，牛角包
culmination ［kʌlmi'neiʃn］n. 顶点，高潮

D

dangerous zone ［'deindʒərəs zəun］危险区域
Danish pastry ［'deiniʃ 'peistri］丹麦包，丹麦酥
degree ［di'griː］（数）度，度数；程度

demand ［di'mɑːnd］ n. 需求

depend on 取决于

deposit ［di'pɔzit］ vt. & vi. 放置，安置

desired star tube 星型挤花嘴

develop ［di'veləp］ vt. 使发展；使发育；开发；培育

difference ［'difərəns］ n. 差别，差异

disintegrated ［dis'intigreitid］ adj. 破裂的，分裂的

dissolve ［di'zɒlv］ vt. 使溶解；使（固态物）溶解为液体

distillation ［disti'leiʃn］ n. 蒸馏（过程）；蒸馏物

divide ［di'vaid］ vt. & vi. 分割，切块

double ［'dʌbl］ adj. 双的；两倍的

double boiler ［dʌbl bɔilə］ n. （美）双层蒸锅

double mixing 两次搅拌

dough ［dəu］ n. 生面团

dough broken down 面筋打断阶段

dough development 面筋开始扩展阶段

dough final 面筋扩展完成阶段

dough letdown ［'let'daʊn］ 面筋搅拌过度

dough stage 主面团阶段

doughnut ［'dəʊ nʌt］ n. 油炸面包圈

dryness ［'drainis］ n. 干燥

due to 由于

E

edge ［edʒ］ n. 边缘

effect ［i'fekt］ n. 反应，效果

egg white foam 蛋白泡沫

egg whites 蛋白

elastic ［ilæstik］ adj. 有弹力的；可伸缩的

emergency ［i'mɜːdʒənsi］ adj. 紧急的，应急的

emergency measure 紧急措施

employ ［im'plɔi］ vt. 使用，利用

emulsified ［i'mʌlsifaid］ adj. 乳化的

enough ［i'nʌf］ adv. 足够地，充足地；adj. 充足的，足够的

essential ［i'senʃl］ adj. 基本的；必要的

essentially ［i'senʃəli］ adv. 本质上，根本上

evaporation ［i'væpə'reiʃn］ n. 蒸发，汽化；消失，发散

evident ['evidənt] *adj.* 明显的，清楚的
exception [ik'sepʃn] *n.* 例外
existing [ig'zistiŋ] *adj.* 目前的；现存的
expandable [ik'spændəbl] *adj.* 可膨胀的，可张开的
experience [ik'spiəriəns] *n.* 经验，体验；经历

F

factory ['fæktəri] *n.* 工厂；复数：factories
fail [feil] *vt. & vi.* 失败，不及格
fall [fɔːl] *vi.* 掉下，跌倒；倒塌，崩溃
farmer bread ['fɑːmə bred] 农夫包
ferment [fə'ment] *vt. & vi.* 使……起发，发酵
filling ['filiŋ] *n.* 填充物；（糕点内的）馅
film [film] *n.* 薄膜
flavor ['fleivə] *n.* 风味；滋味
floor [flɔː(r)] *n.* 地面，地板；楼层
floor time 延续发酵，案台发酵
foam [fəum] *n.* 泡沫
foam cake 乳沫类蛋糕
fold [fəuld] *vt.* 折叠；合拢
form [fɔːm] *vt.* 形成，构成
formation ['fɔːmeiʃn] *n.* 形成
freed [friːd] 释放的
freed gas 释放的气体
French bread [frentʃ bred] 法式面包
fresh [freʃ] *adj.* 新鲜的；淡水的；新的
fruit cake ['fruːt keik] 水果蛋糕

G

gas [gæs] *n.* 气体
gliadin ['glaiədin] *n.* 麦胶蛋白
gluten ['gluːtn] *n.* 面筋；麸质
glutenin ['gluːtənin] *n.* 麦谷蛋白
gram [græm] *n.* （重量单位）克
granulated sugar ['grænjuleitid 'ʃugə(r)] 砂糖

grease ［gri:s］ *vt.* 涂油脂于，用油脂润滑

H

heat penetration 热穿透
hold a party 举办聚会
hole ［həul］ *n* 洞，孔；洞穴
hot dog ［hɔt dɔg］ 热狗
How much… ……多少钱？
hundreds of 好几百

I

I'd like to…我想……
identical to ［ˈaidentikəl tu:］ 与……相同
in solution ［səˈlu:ʃən］ 溶解着；在不断变化中
in spots 有些地方
in turn 相应地；转而
in volume ［ˈvɔlju:m］ 体积上
incorporate ［inˈkɔ:pəreit］ *vt.* 包含；使混合
increase ［inˈkri:s］ *n./vi. & vt.* 增长
indicate ［ˈindikeit］ *vt.* 表明，标示
individual ［ˌindiˈvidʒuəl］ *adj.* 个人的；独特的；个别的
ingredient ［inˈgri:diənt］ *n.* 原料；(混合物的) 组成部分
instantaneous ［ˈinstənˈteiniəs］ *adj.* 瞬间的，即时的
interior pressure 内压
interior ［ˈintiriə］ *adj.* 内部的；室内的
introduction ［ˈintrəˈdʌkʃən］ *n.* 介绍

J

jelly roll 果酱卷

L

lead to [liːd] 通向；导致
liberated [ˈlibəreitid] *adj.* 被解放的；思想解放的，无拘束的
light [lait] *adj.* 光亮的
liquefied [ˈlikwifaid] *adj.* 液化的，溶解的
liquid [ˈlikwid] *adj.* 液体的；清澈的 *n.* 液体，流体
loaf pan 面包听
low fat product 低脂肪产品
low-boiling [ˈləuəbɔiliŋ] *adj.* 低沸点的
low-boiling liquid 低沸点液体

M

mainly [ˈmeinli] *adv.* 大部分地；主要地
make a difference 有影响；起（重要）作用
maturity [məˈtʃuərəti] *n.* 成熟
measure [meʒə(r)] *n.* 测量，测试；措施
mechanical punishment [məˈkænikl ˈpʌniʃmənt] 机械失误后果
meet the sales demand 满足销售需求
migrate into 迁移，移往
milk powder 奶粉
milk solids 奶粉
millions of 成千上万的，无数的
minute gas cell 微小的气室
minute [ˈminit] *adj.* 微小的
mixer [ˈmiksə] *n.* 搅拌机；混合器
moist [ˈmɔistə] *adj.* 潮湿的，微湿的
moisture [ˈmɔistʃə(r)] *n.* 湿度
moisture condition 湿度条件
molasses [məˈlæsiz] *n.* 糖浆；糖蜜
molded [ˈməuldid] *adj.* 成形的
molder [ˈməuldə] *n.* 面团成型机
mousse cake [muːs keik] 慕斯蛋糕/慕思蛋糕/木司蛋糕

N

network ['netwɜːk] *n.* 网；（电视与计算机）网络；网状物
nonfat ['nʌn'fæt] 脱脂的
normal amounts 正常量
normally ['nɔːməli] *adv.* 通常地
no-time dough method 快速发酵法

O

observable [əb'zɜːvəbl] *adj.* 观察得到的，可见的
optimum ['ɔptiməm] *adj.* 最适宜的　*n.* 最佳效果；最适宜条件；（生物学）最适度
optimum condition 最佳状态；最适宜条件
optional ['ɔpʃənl] *adj.* 可选择的；随意的，任意的
orange juice 橙汁
ordinarily ['ɔːdinərili] *adv.* 平常地；通常地；一般地；按说
oven ['ʌvən] *n.* 烤炉
oven rise 炉内膨胀
oven spring 烘焙急胀
over rising ['əuvə'raiziŋ] 过度膨胀
overbake ['əuvə'beik] *vt.* 烘焙过度

P

package ['pækidʒ] *vt.* 把……包成一包
packet ['pækit] *n.* 小包；信息包
pantry ['pæntri] *n.* 餐具室，食品储存室
paper-lined cookie sheet 烘焙纸
partially ['pɑːʃəli] *adv.* 部分地
pastry ['peistri] *n.* 糕点；油酥糕点
pay [pei] *vt. & vi.* 付款
peak of development 面筋扩展最佳状态
penetration [peni'treiʃn] *n.* 渗透，突破
per [pəː, pə] *prep.* 每（表示比率）（尤指数量、价格、时间）
percentage [pə'sentidʒ] *n.* 百分比
perceptible [pə'septəbl] *adj.* 可感觉（感受）到的，可理解的，可认识的
perfect ['pɜːfikt] *adj.* 完美的；正确的；精通的

period ['piəriəd] n. 时期；(一段) 时间；学时
physical ['fizikl] adj. 物理的
pick up 拿起，捡起；取 (给)
place [pleis] vt. 放置
portion ['pɔːʃən] n. 一部分
pound [paund] n. 英镑 (英国的货币单位)；磅 (重量单位)
pour into 慢慢倒入
powdered sugar 糖粉
practice ['præktis] n. 实践
preferable ['prefərəbl] adj. 更好的，更可取的
pre-ferment 第一次发酵
present ['preznt] adj. 现在的，目前的；出席的
pressure ['preʃə(r)] n. 压力；气压
prevailing [pri'veiliŋ] adj. 占优势的，盛行的
prevailing temperature 主温
prevent…from… [pri'vent frɔm] 阻止，防止
price [prais] n. 价格，价钱
procedure [prə'siːdʒə(r)] n. 程序，手续；工序，过程，步骤
process [prəuses] n. 过程；工序；做事方法；工艺流程
processing [prəu'sesiŋ] n. 处理，加工
product ['prɔdʌkt] n. 产品
progressive increase 慢速膨胀
progressive [prə'gresiv] adj. 渐进的
proof [pruːf] vi. 醒发，最后发酵
proof room 醒发室
proper degree 合适的程度
properly ['prɔpəli] adv. 恰当地
proportion [prə'pɔːʃn] n. 比例；部分；面积
protein content ['prəutiːn 'kɔntent] 蛋白质含量
provide [prə'vaid] vt. & vi. 提供，供给，供应
puff [pʌf] n. 泡芙 (奶油空心饼)
puff pastry 松饼，层酥点心，清酥
punch [pʌntʃ] vt. 用拳猛击；翻动面团或翻面
purely physical effect 纯物理反应
purely ['pjuəli] adv. 纯粹地，完全地

Q

quality ['kwɔliti] *n.* 质量，品质。复数：qualities
quantitatively ['kwɔntətetivli] *adv.* 数量上

R

readily ['redili] *adv.* 容易地，无困难地；乐意地
receipt [ri'si:t] *n.* 收据；发票
recipe ['resəpi] *n.* 食谱；处方；配方
recommend [,rekə'mend] *vt.* 推荐，介绍
reduce [ri'dju:s] *vi.* 减少
refer to 指的是，提及，意指
remaining [ri'meiniŋ] *adj.* 剩下的；剩余的
require [ri'kwaiə] *vt. & vi.* 要求；需要
reservation [,rezə'veiʃən] *n.* 预订，预约
rest period 醒发时间
return to 再放回……
rigorously ['rigərəsli] *adv.* 严厉地，残酷地；严密地
ripeness ['raipnəs] *n.* 成熟，老练；成熟度
rough and uneven 粗糙不平
round [raund] *vt. & vi.* 使成圆形
rounder ['raundə] *n.* 面团滚圆机
rye bread 裸麦面包
rye [rai] *n.* 黑麦，黑麦粒

S

salt-free ['sɔ:lt-fri:] 无盐的
sandwich ['sænwidʒ] *n.* 三明治，夹心面包
scale [skeil] *vt.* 测量；称量
scrape [skreip] *vt.* 擦，刮；擦去
seam [si:m] *n.* 接缝，接合处；线缝；裂缝
second fermentation 二次发酵
shape [ʃeip] *n.* 形状；*vi.* 使成形
sheet [ʃi:t] *vi.* 成片展开，擀面，擀薄

shortening ['ʃɔːtniŋ] n. 酥油；雪白奶油
sift [sift] vt. 筛分；精选
sigh [sain] n. 记号；预兆
silky ['silki] adj. 光滑的
size [saiz] n. 大小，尺寸
slack [slæk] adj. 松弛的
slightly ['slaitli] adv. 一点点，轻微
smell [smel] vt. & vi. & link-v. 嗅，闻；闻出
smooth [smuːð] adj. 光滑的；流畅的；柔软的
smooth and mellow 圆润光滑
smooth appearance 光滑表面
so-called adj. 所谓的，号称的
soft foam [sɔft fəum] 湿性发泡
solid ['sɔlid] adj. 固体的；实心的 n. 固体
solubility [sɔljuˈbiləti] n. 溶解度
speed [spiːd] n. 速度；变速器，排挡
sponge [spʌndʒ] n. 面种；海绵；海绵状物
sponge and dough method 二次发酵法，中种发酵法
sponge cake [spʌndʒ keik] 海绵蛋糕
sponge volume 面团的体积
sponge-dough 发酵面团，中种面团，面种（专指面包制作中第一次发酵的面团）
staff mass [stɑːf mæs] 松散的一团
stage [steidʒ] n. 阶段
step [step] n. 步；步骤
sticky ['stiki] adj. 黏性的
stiff [stif] adj. 发泡的；硬的
straight [streit] adv. 直接地
straight dough method 直接发酵法，一次发酵法
stretch [stretʃ] n. 伸展，弹性；vt. 伸展；张开
string [striŋ] n. 线丝，植物纤维
structure ['strʌktʃə(r)] n. 结构；构造；建筑物
subject to ['sʌbdʒikt tuː] 受限于；以……为准；易受……影响
subsequent ['sʌbsikwənt] adj. 接下来的，随后的
successful [səkˈsesfl] adj. 成功的
such as （表示举例）例如，诸如此类的，像……那样的（相当于like 或 for example）
sudden expansion 突然膨胀
sugar ['ʃugə(r)] n. 糖；一块（茶匙等）糖
superior [suːˈpiəriə(r)] adj. （在质量等方面）较好的

surface ['sɜːfis] n. 表面
surface film 表面薄膜
suspend [sə'spend] vt. 悬浮，溶解
sweetness ['swiːtnəs] n. 甜蜜；美妙；芳香
Swiss roll [swis rəul] 瑞士蛋糕卷

T

take place 发生
take up 吸收
temperature ['temprətʃə(r)] n. 温度；体温
tenderness ['tendənis] n. 柔软；温和
texture ['tekstʃə(r)] n. 质地；结构；本质
Thank you for...谢谢你的……
the correct degree 恰当的程度
the fermented sponge 中种面团
the latter 后者
the proofing stage 醒发阶段
the rest 剩下的（人或物）；其他的（人或物）
thermometer [θə'mɒmitə(r)] n. 温度计；体温表
Tiramisu [ˌtirəmi'suː] 提拉米苏
toast bread [təust bred] 吐司面包
tolerant ['tɒlərənt] adj. 可容忍的
transform into 把……转变成……
translucent [trænsˈluːsnt] adj. 半透明的，透亮的，有光泽的
trap [træp] vt. 诱骗；困住
type [taip] n. 类型
typical chiffon cakes formula 有代表性的戚风蛋糕配方

U

ultimate ['ʌltimət] adj. 最后的
uniform ['juːnifɔːm] adj. 均衡的
unique [juˈniːk] adj. 唯一的，仅有的；独特的
unit ['juːnit] n. 单位，单元
up to 达到
utilize ['juːtlaiz] vt. 利用，使用

V

vapor ['veipə] n. 水汽，水蒸气
variation [ˌveəri'eiʃn] n. 不同形状
vinegar ['vinigə(r)] n. 醋
visible ['vizəbl] adj. 看得见的；明显的
visible sign 可见的迹象

W

weaken ['wiːkən] vt. & vi. （使）削弱；（使）变弱；衰减
web [web] n. 蜘蛛网，网状物
weigh [wei] vt. 称……的重量
weight [weit] n. 重量，体重
What kind of... 哪一种……
wheat [wiːt] n. 小麦
whip [wip] v. 抽打；搅拌……直至变稠
whole grain flour 全麦面粉
whole wheat bread 全麦面包
withstand [wið'stænd] vt. 经受，承受；耐得住，禁得起
write down 写下；记下

Y

yeast [jiːst] n. 酵母（菌）；酵母粉，酵母饼，酵母片
yeast ferment 酵母发酵
yogurt ['jəʊgət] n. 酸奶；酸乳酪

附录二
西式面点师职业资格证书理论考试专业英文词汇

工艺过程类：formula（配方）、dough（面团）、sponge（面种/中种面团）、mixing（搅拌）、fermentation（发酵）、divide（分割）、rounding（滚圆）、mold（成形）、final proof（醒发）、bake（烘烤）、temperature（温度）、straight dough method（一次发酵法/直接发酵法）、sponge and dough method（二次发酵法/中种发酵法）、folding（折叠）、cooling（冷却）、sheeting（开酥、擀薄）

面包类：toast bread（吐司面包）、white pan bread（白面包/无盖主食面包）、French bread（法式面包）、Danish pastry（丹麦包）、croissant（牛角酥/牛角包）、farmer bread（农夫包）、whole wheat bread（全麦面包）、multigrain bread（杂粮面包）、sandwiches（三明治）、hot dog（热狗）、doughnuts（油炸面包圈）、rye bread（裸麦面包）、slice of bread（切片面包）

蛋糕类：Black Forest cake（黑森林蛋糕）、Tiramisu（提拉米苏）、mousse cake（慕斯/慕思/木司蛋糕）、cheese cake（芝士蛋糕/奶酪蛋糕）、Swiss roll（瑞士蛋糕卷）、sponge cake（海绵蛋糕）、English muffin（麦芬蛋糕）、chocolate cake（巧克力蛋糕）、chiffon cake（戚风蛋糕）、Angel Food cake（天使蛋糕）、fruit cake（水果蛋糕）、cake decoration（蛋糕装饰）、birthday cake（生日蛋糕）、wedding cake（婚礼蛋糕）

西点类：cookies（曲奇饼）、pie（派/批/馅饼）、apple pie（苹果批）、pumpkin pie（南瓜派）、egg tart（蛋塔/蛋挞）、puff（泡芙/奶油空心饼）、puff pastry（松饼/层酥点心/清酥）、pudding（布甸/布丁）、custard pudding（吉士布甸）

原料类：bread flour（面包专用粉/高筋粉）、cake flour（蛋糕专用粉/低筋粉）、yeast（酵母）、salt（盐）、sugar（糖）、butter（牛油）、margarine（人造奶油）、shortening（起酥油）、milk（牛奶）、milk powder（奶粉）、egg（鸡蛋）、egg white（蛋白）、egg yolk（蛋黄）、emulsifier（乳化剂）、strawberry（草莓）、cherry（樱桃）、peach（桃子）、orange（橙）、mango（芒果）、pine apple（菠萝）、kiwi（奇异果）、star fruit（杨桃）、honeydew-melon（蜜瓜）、red cherry（车厘子）、pear（梨）、gelatin（明胶）、chocolate（巧克力）、white chocolate（白巧克力）、icing sugar（糖粉）、candied sugar（脆糖）、soda（苏打粉）、cream of tartar（塔塔粉）、

baking powder（发粉/发酵粉/泡打粉）、custard（吉士粉）、jelly（果冻）、raisin（葡萄干）

机器类：mixer（搅拌机）、divider（面团分割机）、rounder（面团滚圆机）、molder（成型机）、final proofer（醒发箱）、oven（烤炉）、slider（面包切片机）、sheeter（开酥机）、package/packer（包装机）、cooler（冷却器）

工具类：scale（秤）、plastic scraper（塑料刮刀/塑料刮板）、roller（擀面杖）、piping bag（裱花袋）、baking pan（烤盘）

其他：baker（面包师）、bakery（面包店）、cake shop（蛋糕店）

附录 三
拓展阅读参考译文

拓展阅读1：上海的一间面包店

我们的面包店于2000年在上海国际贸易中心盛大开业。现在我们有1 200平方米的厂房。在这里，我们提供您所需的任何大小、形状或味道的产品：烘焙食品、巧克力、糕点、饼干、冰淇淋杯和挞类等。平价的烘后冷冻成品——烘烤后的成品经急速冷冻、冷冻和即食成品可用于每天的生产。随着生产经验的积累，我们已经能够应对大批量烘焙产品的生产，如2010年上海世博会、上海网球大师赛、F1赛车、上海高尔夫球公开赛等的烘焙产品供应。

我们拥有一支40名面包师的队伍，还有一个德国的和一个荷兰的面包大师，使用进口原料生产多种类的烘焙产品。欧洲圣诞节、复活节的时令特色和中国新年的特点，都会呈现在我们的产品中。

我们的客户大多是上海、苏州、无锡的四星级和五星级酒店、国际餐厅和许多连锁超市。

巴斯面包有限公司现在是上海式最大和最古老的欧洲式面包店，是许多国际知名城市连锁酒店、连锁超市和连锁餐厅的首选供应商。

迄今为止，我们取得QS和ISO 22000的认证已经6年了。

巴斯面包，您对高品质烘焙产品的首选！

（陈明瞭译）

拓展阅读2：庆典蛋糕

当谈到生日、毕业等庆典的时候，蛋糕总是大家感兴趣的一个关键点。生日聚会没有美味蛋糕的话真的会不一样。

您绝不会因为太老了而不能吃蛋糕。那为什么不到附近我们这个蛋糕店（The Cake Shop）定制一个美味的蛋糕呢？我们从事这一行业已经25年了，我们能为您设计、制作任何你想要的蛋糕。您的想法会被如实地体现。

我们已经制作过许多不同款式、不同大小和不同设计的庆典蛋糕。所以无论是儿童生日蛋糕、毕业蛋糕、洗礼蛋糕还是女性蛋糕，只要是你想要的，我们都能满足您。

在英格兰中部大部分地区，包括牛津郡、沃里克郡、白金汉郡、威尔特郡、格罗斯特郡和伯克郡，我们提供送货服务。而且，您可以从我们任何一个蛋糕店提取结婚蛋糕。

一个极为独特的事件值得一个独特的蛋糕。所以今天就上网看看，下一步就去我们的蛋糕店定制一个庆典蛋糕吧！

(章佳妮译)

拓展阅读3：三明治

英国人通常会在午餐时间吃三明治，当然也有在别的时间吃的。三明治由两片面包中间夹着馅料，一般面包上会涂上黄油或者蛋黄酱，而中间的馅料通常是肉或者奶酪，再放些生菜。

然而，三明治的种类有几百种，而每一种都有其独特的风味。在英国，吃得最多、最著名的几种三明治包括：BLT（也就是培根、生菜加西红柿，一般还要加上蛋黄酱）、庄稼汉汉堡（源自农场工人传统的做法，配料有切达奶酪、泡菜和沙拉酱），还有金枪鱼和鸡蛋三明治。

因为三明治吃起来方便快捷，因此不管是什么种类的三明治都很受欢迎。事实上，英国人每年要吃掉28亿个三明治，对于人口仅6 000万的国家来说，这很了不起！今天的英国人都在吃三明治，但其实以前可不是这样的。不可思议的是，如今那么不起眼的三明治，原来以前只是英国富豪们的小点心，三明治还有一段有趣而又幽默的历史。

1762年，sandwich这个单词第一次出现在英国作家爱德华·吉本斯的日记上，他当时回忆看见全国首富在吃一种在面包片中夹一小片冷肉的东西，吉本斯觉得这样的吃法对这么高贵的人来说很不合适。

这种小吃以桑威奇（与三明治同音）伯爵四世的名字命名（伯爵是富有的贵族，拥有大片土地，也拥有一定政治权力）。桑威奇惯赌，痴迷赌博甚至到了饭也不想吃的地步，为了不中断赌博，桑威奇伯爵就让赌博俱乐部的厨子为他准备两片面包，中间加些牛肉来吃，这样他就可以空出一只手来玩牌，再也不用为了吃肉弄脏两只手了。

其他人看到他这么吃，也都开始点菜："来份跟桑威奇一样的！"于是三明治就这么诞生了，一开始它只是英国有钱人家的小吃而已，后来迅速流行起来，成为各处都可吃到的便捷食品。

拓展拓展4：面包知多少？

面包是西方菜单上不可或缺的一种食物。有时你会看到菜单上有很多种面包，这时选择合适的面包来搭配你的主菜就是一项不简单的任务了。如果你在国外，了解不同种类的面包和它们的正确吃法是十分有用的。

面包基本上分三种：

1. 发酵面包

发酵面包中含有气孔，面团中的酵母能使面团在烘烤期间膨胀发大而成为面包的样子。例如，软式面包中的葡萄干面包和全麦面包就是发酵面包。比萨饼和汉堡也属于发酵面包。

2. 快速面包

制作快速面包所花的时间比发酵面包要少。它们的发酵效果是由发酵粉，而不是酵母引起的。例如，玉米面包、甜甜圈、麦芬和薄饼都属于快速面包。

3. 平面包

这种面包就像它的名字一样，是平的。平面包包括墨西哥玉米饼皮、印度薄饼和中东皮塔饼。通常，它们都有用来填充馅和调味料的空隙。

吃面包的正确方法：

在正规的西餐厅，面包盘与黄油刀通常放在你的左手边。吃面包的时候，你应该把面包撕成小块。用刀切或直接咬都是不礼貌的。你只能在要吃的面包块上涂黄油（盘子里的其他面包不用）。面包屑掉在盘子里或桌子上都是可以的，服务生会负责清理。

拓展阅读5：面包是什么？

在这个国家，为了大家的利益，面包食品达到高标准的干净和卫生。

面包也许是我们饮食中最重要的东西，它经常被称为"生活的本质"。为了使你更好地了解面粉和面包的好处，一项政府调查表明面粉和面包比任何其他基本食品为我们提供了更多的能量值，更多的蛋白质、铁，更多的烟酸和维生素B1。面包以许多有趣的形状和味道呈现在我们面前：从历史悠久的"山寨"面包，到美味的维也纳卷。如今，切片和包装的面包是最受欢迎的面包。面包做的三明治是理想的野餐食品，也是工人们理想的午餐。然而，面包也有一个很不好的地方。如果你喜欢面包上面有一层漂亮的金色面包皮的话，就不要买包装好的那些。生活中最美好的事情之一就是放学或下班回家饿了，能吃上刚做出来的新鲜酥皮的农舍面包或科堡包。

面包在我们的生活中如此重要，所以它应该被更加重视和充分享用。在你的城市，也许就有许多面包师。找找看你通常喜欢吃哪个面包师的面包。除了普通的白面包、全麦和小麦粉面包外，还有许多被面包师称为"梦幻型"的其他面包，如"麦芽"面包、葡萄干面包、牛奶面包（含有奶粉）和各种茶面包。还有香料面包，比如姜面包，但它实际上应该是蛋糕。而在荷兰它总是出现在早餐桌上。

时间向前推进，在许多工业领域中全机械化被提上议事日程，正如你所看到的那样，烘焙业在迅速机械化。目前只需要45分钟就可以在炉中烤好一个普通的面包。但是，人们已经在开发能烤面包的电子设备，利用高频加热，三分钟就够了。面包烤得如此之快，却没有时间形成面包皮——这样的产品就没有吸引力了。然而，在国际突发事件中，如大地震、洪水等情况下，它会有很大的用处。当有数千人急需食物的时候，邻近的国家可以在很短的时间内生产大批量的面包并将这些面包送到受灾地区。你有没有想过一年你要吃多少面包以及肉、土豆、蔬菜等？你可能要吃超过100公斤或近两倍于自己体重的东西。

你可以确定一件事——面包是我们可能得到的最好的食物之一。事实上，可以毫不夸张地说，没有它，我们就不能过日子了。有很多种食品和奢侈品，如冰淇淋或糖果，没有它们我们也照样可以把日子过得很好，有了它会更健康。均衡的饮食使人身体强壮，精神爽朗，因此身体必须要吸收生命的本质——高质量的面包。

（章佳妮译）

拓展阅读 6：关于蛋糕

蛋糕是一种面包或面包类食品。它的现代形式是甜的烘焙点心，它最古老的形式是炸面包和芝士蛋糕，通常是圆形。要区分一种食物是面包、蛋糕，还是糕点类，是很难的。

现代蛋糕，尤其是多层蛋糕，通常含有面粉、糖、鸡蛋和黄油或油，有些品种还需要液体（通常是牛奶或水）和膨松剂（如酵母或发酵粉）。一些可增加风味的原料也经常被添加进去，如果酱、坚果、干果或果脯，或浸膏剂和众多的替换物。蛋糕中通常会有水果蜜饯、甜点酱汁（像奶油）、冻奶油或其他糖霜，或用杏仁膏进行装饰，或用裱花围边，或用蜜饯来装饰。

蛋糕通常是庆典用餐选择的甜点，尤其是婚礼、周年纪念日和生日。有无数的蛋糕配方：有的蛋糕配方像面包，有的蛋糕配方营养价值高而精细，有许多蛋糕配方历史悠久。蛋糕的制作已不再是一个复杂的过程，而曾经要花费大量劳力的蛋糕制作（特别是搅拌鸡蛋泡沫），由于烘烤设备及说明书进行了简化，所以，即使是最业余的厨师也可以烤制蛋糕。

（章佳妮译）

拓展阅读 7：汉堡包

由两片面包之间夹一块碎肉饼而成的汉堡包，最初是在 1900 年由美国康涅狄格州纽黑文的 Louis' Lunch 饭店老板，一位丹麦移民，路易斯·拉森发明的。关于汉堡包的由来，Charlie Nagreen，Frank and Charles Menches，Oscar Weber Bilby 和 Fletcher David 一直都有不同的说法。White Castle 追溯汉堡包的起源到德国的汉堡市，认为是由 Otto Kuase 发明的。然而在 1904 年圣路易斯世界博览会上《纽约论坛报》莫名地认为汉堡包是"收费高速公路上食品小贩的发明"，那时它得到了全国人民的认同。自从汉堡包问世以来，对于谁是汉堡包发明者这个问题的争端就没有结束过，至今没有令人信服的说法。

国会图书馆已正式承认是康涅狄格州纽黑文 Louis' Lunch 饭店的路易斯·拉森，这位马车午餐店老板于 1900 年在美国出售了第一个汉堡包和牛排三明治。《纽约杂志》说，"这种食物没有名字，直到几年后才由从汉堡来的几个吵闹水手将它命名为汉堡包。"《纽约杂志》还评论说这种观点是令人认可的。当年，一个顾客点了热的快餐，路易斯的牛排刚好用完了，于是他拿绞牛碎肉做了一个饼，烤后放在两片面包之间。然而一些评论家，如 Josh Ozersky，《纽约杂志》的美食编辑，认为这个三明治不是汉堡包，因为面包是烤过的。

汉堡包是快餐店的特色。大型快餐店供应的汉堡包通常由工厂大规模生产，冷冻后送到需要它的地方。这些汉堡包薄且有均匀的厚度，不同于传统在家做的或在传统的餐馆做的美式汉堡包，美式汉堡包比较厚，它是用绞牛肉手工制作的。大多数美国汉堡包是圆的，但一些快餐连锁店，如温迪快餐连锁店就出售方形的汉堡包。快餐店供应的汉堡包通常在一个平面上烤，但一些公司，如汉堡王，用的是气体火焰烧烤工序。在传统的美国餐馆，汉堡包可以点"半熟的"，但考虑到食品安全通常都是七分熟或熟透。快餐店通常不提供选择。

麦当劳快餐连锁店出售巨无霸汉堡包。这是世界上最畅销的汉堡包之一，在美国，估计每年可售出五亿五千万个。其他主要的快餐连锁店包括 Burger King（在澳大利亚也被称为 Hungry Jack's），A&W，Culver's，Whataburger，Carl's Jr./Hardee's chain，Wendy's（以方形饼而闻名），Jack in the Box，Cook Out，Harvey's，Shake Shack，In-N-Out Burger，Five Guys，Fatburger，Vera's，Burgerville，Back Yard Burgers，Lick's Homeburger，Roy Rogers，Smashburger，Taco Bell 和 Sonic，也是主要销售汉堡包的。而 Fuddruckers 和 Red Robin 是专门经营中档"餐厅风格"品种的汉堡包连锁店。

一些北美的饮食机构用独特的方法做汉堡包，完全不同于快餐店的做法。它们采用高档材料如牛腰肉或其他肉排以及各种不同的奶酪、配料、调味料。例如，由知名厨师和食品网络明星博比·弗莱建立的博比汉堡皇宫连锁店就是其中之一。

汉堡包常作为快餐食品、野餐或聚会食品，也常用于户外烧烤。

（章佳妮译）

拓展阅读8：面粉的特殊分类

面包粉：含13.5%～14%的面筋。

蛋糕粉：面筋含量低。

全麦粉：全麦粉是由整个麦粒研磨而成，含有胚乳、胚芽、麸。然而我在中国买的全麦粉似乎少了"全"字，我猜想麦粒被过分研磨了，有部分麸被磨掉了，颜色不再那么暗，从而增强面粉口感和结构。我猜想它更像是T90或T110的法国面粉而不是亚瑟王全麦面粉。

饺子粉：一种相对高筋面粉（大约11%）。如果你找不到面包粉或高筋粉的话，饺子粉也是一个不错的选择，因为在很多超市里都有卖（事实上，这是我见过的唯一一种到处都能买到的面粉）。

油条粉：这种面粉含有几种化学添加剂，用于产生有气孔、有嚼头、海绵状的油条，一种深度油炸的面食小吃。我没有试过这种面粉，也不打算尝试。

自发粉：一种中筋度、多功能面粉，经常会添加烘焙苏打（小苏打）、酸式盐，有时还会加酸性磷酸钙（这种添加剂也会出现在烘焙面粉里）。

澄粉：是一种完全去除面筋的面粉。它实质是一种麦淀粉，由于其显著的耐嚼性和黏性，它用来制作有黏性的年糕和糯米团。

麦心粉：由麦粒的胚芽磨制而成。

（龙小清译）

拓展阅读9：面包的历史

从最早时期开始，面包就以某种形式作为人类主要食物之一而存在。

制作面包是世界上最古老的手艺之一，条形和卷状面包被发现于古埃及的墓穴中，在大英博物馆里的埃及陈列馆你就能看到真正的条形面包，其制作并烘烤于5 000多年以前。同时展出的还有成熟于法老时期的那个古老夏天的麦粒。在8 000多年前活跃的人类定居点的深坑内发现了麦粒，发酵和未发酵的面包《圣经》里都多次提到。甚至在人们争论是白面粉好还是褐色面粉好的时期里,古希腊人和古罗马人已经把面包当成主要食物了。

再往回追述,在石器时代,人们用石头碾碎的大麦和麦粒粉来制作坚硬的蛋糕。人们发现一个用来磨谷物的石磨,它被认为有 7 500 年的历史了。拥有种植和收获谷类植物的能力可能是引导人类社区定居的一个主要原因,而不再是过着到处打猎和放牧的流浪生活。

根据植物学家的观点,小麦、燕麦、大麦和其他谷类都属于草科,然而至今人们还未发现进化成为现在我们所认为的小麦的野草。就像大多数野草,谷类的花也是雌雄同株。为了确保幼株发芽,谷类植物就储存一定量的营养食物,正是这种物质的储存才让人类发现了大量的食物。

(龙小清译)

拓展阅读 10:蛋糕的历史

蛋糕是由不同的成分制作而成的,包括精制面粉、起酥油、糖、蛋、牛奶、膨胀剂、香精。毫不夸张地说,蛋糕的配方多达数千种(有些类似面包,有些则是细腻、多油),有些有几百年的历史。蛋糕的制作不再是一个复杂的过程了。

制作蛋糕的器皿和指引已经如此完善、简单,以至于非专业的厨师也能轻易地成为专家级的烘焙师。这里有五种基本类型的蛋糕,它们的区别取决于发酵的物质。

世界上最早出现蛋糕是在人们发现面粉后不久。在中世纪英国,用文字描述的蛋糕并不是我们现在传统意义上的蛋糕,相对于面包是没有糖的面粉食物的描述,蛋糕则被描述成面粉甜品。

当初面包和蛋糕是两个可以交替使用的词汇,而蛋糕被用来指较小一点的面包。最早的例子是发现于新石器时代村落遗址,在那里考古学家发现由磨碎的谷物制作的简单蛋糕,它们被和湿、压实,最后可能是在滚烫的石头上烘烤而成的。这种早期的蛋糕今天可以被认为是大麦蛋糕,尽管现在我们更多地认为是一种饼干或曲奇。

希腊人称蛋糕为"Plakous",由单词"flat"演化而来,这类蛋糕通常是由坚果和蜂蜜制作的,希腊人还有一种扁平的重油蛋糕,被称为"satura"。

在罗马时期,蛋糕的名字成了"placenta"(由希腊词汇演化而来),也被罗马人称为"libum",最初被用作供神祭品。Placenta 更像是一种奶酪蛋糕,在一种糕点的基础上烘烤得来,有时置于一个糕点盒内。

"面包"和"蛋糕"这两个词随着时间推移变得可以相互代替,这两个词本身属于盎格鲁—撒克逊时期沿袭的英文单词,很有可能"蛋糕"这个词用来表示较小的面包,通常"蛋糕"是用于特殊场合,因为它们是用当时可获得的最好和最贵的原材料制作而成,你越富裕,就越有可能频繁地消费蛋糕。

18 世纪中叶,在蛋糕制作中酵母被错用为一种膨松剂以取代搅匀的蛋液。一旦大量空气被打入,面团就倒入模具,经常会有复杂的造型,但有时非常简单,就是两个锡制的圈圈,放置在羊皮纸上,再铺上烘焙油纸。正是在这些蛋糕圈圈的基础上才发展出我们现代的蛋糕模。

早期美国东部沿海的厨师认为蛋糕是幸福生活的象征,全国每个地区都有他们自己的喜好。

19世纪早期，由于工业革命，大量的制造生产和铁路的开通使用，烘焙原材料变得越来越便宜且随时可以获得，现代的膨胀剂，比如烘焙苏打和泡打粉也已经发明了。

（龙小清译）

拓展阅读11：10种有益健康的面包

在百货商店，出售面包的区域深受人们欢迎，它也是一个让人相当纠结的地方。面包的原料是一个相当重要的方面，一定要选那种百分百是全谷物的面包。谷物是纤维的一大来源，能防止患心脏病和结肠癌。纤维也能让人长时间有饱腹感，这样我们每天就可以少吃些东西，可以维持体重。以下列出了一些有益健康的面包种类。

1. 亚麻籽面包：加有亚麻籽的面包不仅吃起来松脆可口，亚麻籽里的胆固醇成分还可以保持心脏健康。亚麻籽里还有ω-3脂肪酸，因此毫无疑问，素食主义者可以食用这种面包。

2. 皮塔面包：皮塔面包通常被认为是一种没营养的面包。但是它是地中海沿岸人群的食物，因有益于心脏健康而闻名于世。据食用过这类食物的人说，食用它可以减少心脏紊乱的概率。

3. 无麸质面包：麸质是一种蛋白质，小麦、大麦以及黑麦中都含有麸质。无麸质面包通常由谷物制成，而不是由小麦制成。全谷物无麸质面包可以用糙米粉来制作。

4. 粗裸麦面包：粗裸麦面包基本上就是黑麦面包的一种，除了黑麦，还会加入豆类和全谷类。这就是粗裸麦面包被称为最健康的面包种类的原因。

5. 核桃面包：核桃是世界上营养最丰富的坚果之一。核桃富含ω-3脂肪酸。因此，核桃面包含有很多对心脏有益的胆固醇。

6. 燕麦面包：燕麦面包是一种可供我们食用的最健康的面包。燕麦对心脏健康非常有益，因为它们可以将有害的胆固醇加以吸收。由燕麦制成的面包含有很丰富的纤维，同样也有益于心脏的健康。

7. 黑麦面包：黑麦面包是一种市场上可以买到的最健康的面包。但一定要确认面包里含有黑麦，同样面包里也有黑麦的味道。

8. 杂粮面包：杂粮面包是由很多种谷物混合到一起制成的，如小麦、黑麦、小米等。这也是它富含营养纤维的原因。同样，它也对保护心脏有好处。

9. 黑面包：黑面包通常是由小麦和大豆混合制成的，这也造就了它黑黑的颜色。黑面包也有益于心脏的健康，主要是因为它是低脂面包且无麸质。

10. 全麦面包：全麦面包是由未经加工的小麦制成的。这就是它有益于我们的身体健康、富含纤维、对心脏有益的原因。全麦、富含纤维，面包的上佳选择。

希望本文能够帮助读者选到合适的面包种类，保持心脏健康。

拓展阅读12：烤炉

烤炉是一个为面包产品加热的装置，在商业面包生产中这是一个非常重要的生产设备。烤炉有一个热源装置、一个通常配有蒸汽加湿器的烤箱、一个有装卸面包设备的传送带、一个烘焙过程自动控制系统、一个热气回收的装置。烤炉内的平均温度达到200℃～

300℃，相对湿度15%～70%。至于工作过程和设备的性质，烘焙烤炉和甜品烤炉相似。

  根据操作方法的区别，烤炉分为间歇式烤炉和连续式烤炉。按烤箱设计结构不同，烤炉分为传统的箱式炉和隧道炉（单层炉和多层炉）两大类。对于传统的烤炉来说，面团的入炉和成品的出炉操作都在同一边，而隧道烤炉的这两种操作是在两面进行。烤炉还可以根据炉膛的类型分类，链式传送带上安装支架炉膛，盘子安装在链式传送带上形成一个连续的水平炉膛；格子框架炉膛是以皮带传送带，圆盘围绕一个纵轴旋转，一个环形或一个拉模板，或固定的单独炉膛。烤炉还可以根据热源分类，它们有提供高压饱和水蒸气或过热蒸汽的管道部分，燃料和循环热气的燃烧热量穿过隧道，可以是平的或管道的隧道，或用电炉、煤气灶、红外辐射灯直接对烤箱进行加热，有些烤炉则使用多种加热方式进行烘烤。

  烤炉在不同程度地机械化，带有传送带炉膛的自动烤炉，传送带的运送、加热的条件、蒸汽的供应都是自动控制，面团的入炉和成品的出炉也都机械化或自动化了，这些都是由一个自动装置确保燃料燃烧的安全。具有传送带的机械炉，传送带是被机械化控制进行成品的出炉，这些也是由一个自动装置确保燃料燃烧的安全。其他机械烤炉有烤盘或延长炉膛，它们正在被更现代化的设计所取代，而非机械化的烤炉现在仅在小型企业使用。

  烤炉的生产率取决于设计的先进程度，有的产量可以达到每天100吨～120吨。

（龙小清译）

参考资料

1. E. J. Pyler. *Baking Science and Technology*. （3rd ed） Chicago：Siebel Publishing Company，1982.

2. Sylvia M. J. Enkins. *Bakery Technology*. Toronto：Lester and Orpen Limited，1975.

3. *The History of Sourdough*，http：//www. kitchenproject. com/history/sourdough. htm.

4. *The Great Soviet Encyclopedia*，http：//encyclopedia2. thefreedictionary. com/Baking + Oven，1979.

5. *Baking Bread in China：A Flour Glossary*，http：//www. hawberry. net/baking-bread-china-guide-ingredients-supplies/ flour-guide/.

6. *History of Bread*，http：//www. botham. co. uk/bread/history1. htm.

7. Larsen Linda. *Bread Ingregdients*，Bread 1o1，http：//busycooks. about. com/od/howtobake/a/brea.

8. Hamburger，http：//en. wikipedia. org/wiki/Hamburger.

9. Cake，http：//en. wikipedia. org/wiki/Cake.

10. *The Story Behind a Loaf of Bread*，http：//www. botham. co. uk/bread/bread1. htm.

11. http：//www. the-cakeshop. co. uk/newshop/wedding-cakes. asp.

12. http：//whatscookingamerica. net/History/CakeHistory. htm.

13. 英国总领事馆文化教育处资料。

14. 《基础英语》模块第一、二册. 北京：外语教学与研究出版社.

15. 英王面包店 A-1 Bakery，http：//www. openrice. com. cn/hongkong/ canting/detail/14867.

16. 工厂图，Buns of Bread in the Factory，http：//www. dreamstime. com/royalty – free – stock – images – buns – bread – factory – image19331419.

17. Danish Style Cake Shop & Delicatessen，http：//www. blogto. com/bakery/danishstyle – bakeshop.

图书在版编目（CIP）数据

烘焙专业英语 = English for Baking/陈明瞭主编．—广州：暨南大学出版社，2014.6
（2023.7重印）
（食品生物工艺专业改革创新教材系列）
ISBN 978-7-5668-0991-9

Ⅰ.①烘…　Ⅱ.①陈…　Ⅲ.①烘焙—英语—高等学校—教材　Ⅳ.①H31

中国版本图书馆CIP数据核字（2014）第069206号

烘焙专业英语
HONGBEI ZHUANYE YINGYU
主　编：陈明瞭

出 版 人：张晋升
策划编辑：张仲玲
责任编辑：王嘉涵
责任校对：杨柳婷
责任印制：周一丹　郑玉婷

出版发行：暨南大学出版社（511443）
电　　话：总编室（8620）37332601
　　　　　营销部（8620）37332680　37332681　37332682　37332683
传　　真：（8620）37332660（办公室）　37332684（营销部）
网　　址：http://www.jnupress.com
排　　版：广州市新晨文化发展有限公司
印　　刷：广东信源文化科技有限公司
开　　本：787mm×1092mm　1/16
印　　张：7.5
字　　数：188千
版　　次：2014年6月第1版
印　　次：2023年7月第8次
印　　数：12501—14000册
定　　价：38.00元

（暨大版图书如有印装质量问题，请与出版社总编室联系调换）